5.90

# The Structure of the
# REAL NUMBER SYSTEM

# The Structure of the
# REAL NUMBER SYSTEM

LEON W. COHEN
*Professor Emeritus of Mathematics, University of Maryland*

and

GERTRUDE EHRLICH
*Professor of Mathematics, University of Maryland*

Robert E. Krieger Publishing Company, Inc.
Huntington, New York 1977

*Original edition 1963*
*Revised edition 1977, with extensive changes*

*Printed and Published by*
ROBERT E. KRIEGER PUBLISHING, INC.
*645 New York Avenue*
*Huntington, New York 11743*



*Printed in the United States of America*

Cohen, Leon Warren, 1903–
The structure of the real number system.

Bibliography: p.
Includes index.
1. Numbers, Real. I. Ehrlich, Gertrude, joint author. II. Title.
QA241.C67 1976 512′.7 76–7512
ISBN 0-88275-396-7

# PREFACE TO THE REVISED EDITION

In this revised edition, each section of the text has been reorganized, with all exercises appearing at the end of the section. All theorems, definitions and exercises have been renumbered. (Example: the third theorem of the fourth section of Chapter 2 is referred to as Theorem 2, 4.3, or simply as Theorem 4.3 in case the reference occurs within Chapter 2). New exercises have been added throughout the book and old exercises have been revised and, in some cases, supplied with "hints."

Numerous changes have been made in the text, including major revisions in the treatment of sets, equivalence relations, functions and counting. Groupoids and semigroups have been eliminated. The treatment of functions has been modernized and brought into conformity with contemporary language. One of the authors gladly acknowledges that his collaborator bore the major share of the burden of this revision.

LEON W. COHEN
GERTRUDE EHRLICH

# PREFACE TO THE FIRST EDITION

The present course of mathematical education in school and college introduces a student rather casually to the various properties of integers, rational numbers, and real numbers as they are needed for arithmetic, algebra, geometry, and calculus. Sometimes the relations among these mathematical structures are sketched as indications of things to come in mathematics. When the student begins graduate study he finds that he is expected to be familiar with the structure of the real number system. If his graduate courses deal explicitly with this system at all, they usually do so in terms of a brief summary. It is then assumed that the student has a usable knowledge of the intricately interwoven properties of set theory, algebra and topology which characterize the system of real numbers. Actually, this assumption is frequently false and a gap is left in the student's knowledge which, if not filled, hinders his development.

This little book, it is hoped, will help fill the gap as a text either for independent study or for a semester course. The exercises in each chapter extend and illustrate the main line of the theory. Some of the exercises are referred to in proofs of theorems. In such cases, we have identified them thus: ●

The authors are indebted to many colleagues for helpful criticism while the manuscript was being used as the basis for a course. Mrs. Mary Gray provided indispensable and sympathetic assistance in typing the several revisions of the manuscript.

LEON W. COHEN
GERTRUDE EHRLICH

*June 1963*
*College Park, Maryland*

# CONTENTS

# CHAPTER 0

# INTRODUCTION

**1. PRELIMINARIES.** It may just as well be understood that one cannot begin a discussion of mathematics, or even of a portion of mathematics, at the beginning. One may look for such a beginning by setting down, in the conventional manner, undefined terms, unproved propositions, and a proof scheme. The hope that a tidy mathematics might be deducible in this way is implicit in Euclid's geometry. The search for an axiomatic system which might provide a foundation for all of mathematics and a proof of its own consistency continued through the first third of this century.

But there were essential difficulties. The operation of a formal mathematical system requires some instructions stated in a language outside the system. It has been known since 1932 that, even if such instructions are admitted as understood, no system adequate for the most familiar part of mathematics, the arithmetic of the whole numbers, can be proved to be consistent. Thus, one must be content to begin somewhere in the middle, using the axiomatic method, in spite of its limitations, as a means of organizing portions of mathematical knowledge.

The real number system has strong claims to a central position in mathematics. It is the point of departure for the vast field of mathematics called "analysis." Together with its subsystems, the real number system provides models and techniques for much of set theory, algebra, geometry, and topology. It was the critical study of the real numbers, requiring the re-examination and reconstruction of logic, which finally ended the hope of finding a beginning. Because the germ of so much mathematics lies concealed within the real numbers, we hope that this book will serve not only as an introduction but as an invitation to mathematics.

As concepts are defined in the following pages, examples will be given either to illustrate or to motivate them. The examples will be drawn from familiar mathematical discourse as well as from nonmathematical experience. Strictly, the only examples of mathematics are concepts and relations expressed in the terms, axioms, and theorems of mathematics. In fact, however, these constituents of mathematics are usually abstracted from experience, and experience includes what may be called familiar mathematics. There seems to be an historical cycle in which the strict mathematics of one epoch becomes the familiar mathematics on the basis of which a stricter, more inclusive, mathematics is later formalized. From this point of view it is not strange that the familiar Cartesian plane serves as an example motivating the definition of a Cartesian product in terms of which the Cartesian plane is formally defined. Put briefly, the problem as to which of "abstraction," "experience," is "chicken," "egg" is not resolved.

**2. SET THEORY.** The fundamental concepts of mathematics may be expressed in the terminology of set theory. We collect in this chapter some facts of the theory of sets which will be used later.*

The terms "set" and "element of a set" will not be defined. Although the mathematical objects treated in this book will all be sets, we shall, in some contexts, explicitly refer to a set as a "set of sets." Sets will usually be denoted by capital letters; but when a set occurs as an element of another set, we may denote it by a small letter. We write "$a \in A$" if $a$ is an element of $A$, "$a \notin A$" if $a$ is not an element of $A$, and denote by the symbol "$\{a, b, c, \ldots\}$" the set whose elements are $a, b, c, \ldots$. (If $A \in B$ and $B \in C$, it does not follow that $A \in C$. For example, if the United Nations is a set whose elements are the member nations, and a nation is a set whose elements are its citizens, then Dieudonné is an element of France, France is an element of the United Nations, but Dieudonné is not an element of the United Nations.)

AXIOM OF EXISTENCE *There is a set.*

AXIOM OF IDENTITY *If $A$, $B$, are sets then $A$ and $B$ are the same set if and only if every element of $A$ is an element of $B$, and every element of $B$ is an element of $A$.*

We write "$A = B$" if $A$ and $B$ are the same set, "$A \neq B$" if $A$ and $B$ are not the same set. At times, we shall find it convenient to use the symbol for an element of a set repeatedly. Thus, for example, the symbols "$\{a, a, a, b\}$," "$\{a, a, b\}$," "$\{a, b\}$" will all denote the same set.

*A more extensive account of the subject at an appropriate level is found in *Native Set Theory*, by P. Halmos (D. Van Nostrand Co., Inc., Princeton, N.J., 1960), from which we take some of our undefined terms, axioms, and definitions.

DEFINITION 0.1 Set $B$ is a *subset* of set $A$ if $b \in A$ for all elements $b$ of $B$. If $B$ is a subset of $A$, we write "$B \subset A$" or "$A \supset B$." If $B \subset A$, and $B \neq A$, then $B$ is a *proper subset* of $A$.

For example, the set of all even integers is a proper subset of the set of all integers. The set of all non-mathematicians is a proper subset of the set of all human beings. The set consisting of Great Britain and $\sqrt{2}$ is a proper subset of the set consisting of Great Britain, $\sqrt{2}$, and the moon.

THEOREM 2.1 1) *If $A$ is a set, then $A \subset A$.*
2) *If $A$, $B$ are sets, then $A = B$ if and only if $A \subset B$ and $B \subset A$.*
3) *If $A$, $B$, $C$ are sets such that $A \subset B$ and $B \subset C$, then $A \subset C$.*

*We leave the proof as an exercise (Exercise 2.1).*

We observe that, in the examples of subsets given above, each subset is determined by a condition imposed on the elements of the original set. The subset of all even integers is determined by the condition "$x$ is even" imposed on $x$, where $x$ is an integer; the subset of all non-mathematicians by the condition "$x$ is a non-mathematician" imposed on $x$, where $x$ is a human being; the subset consisting of Great Britain and $\sqrt{2}$ by the condition "$x$ is Great Britain or $\sqrt{2}$" imposed on $x$, where $x$ is Great Britain, $\sqrt{2}$, or the moon.

AXIOM OF SPECIFICATION *If $A$ is a set and $Q(x)$ is a condition, then there is a subset $B$ of $A$ whose elements are exactly those elements $x$ of $A$ for which the condition $Q(x)$* holds.

The subset $B$ is said to be *determined* (or *specified*) by the condition $Q(x)$. By the Axiom of Identity, if $B$ and $B'$ are subsets of $A$ determined by a condition $Q(x)$, then $B = B'$. We shall say: "a unique subset $B$ of $A$ is determined by $Q(x)$", and write

$$B = \{x \in A \mid Q(x)\}.$$

THEOREM 2.2 *There is a set which has no elements.*

PROOF: By the Axiom of Existence, there is a set. Let $A$ be a set, and let $Q(x)$ be the condition: "$x \notin A$." By the Axiom of Specification, there is a subset $E$ of $A$ consisting of all elements $x$ of $A$ which satisfy the condition "$x \notin A$." Since no element of $A$ satisfies this condition, the set $E$ has no elements.

A set which has no elements is called an *empty set.*

THEOREM 2.3 *If $E$ and $E'$ are empty sets, then $E = E'$.*

PROOF: Suppose $E \neq E'$. Then one of the following statements must be true:

(1) There is an element $x \in E$ such that $x \notin E'$.
(2) There is an element $x \in E'$ such that $x \notin E$.

But both of these statements are false, since neither $E$ nor $E'$ has any elements. It follows that $E = E'$.

In view of Theorem 2.3, we shall speak of "*the* empty set." We denote this set by "$\phi$."

*Remark:* For any set $A$, $\phi \subset A$ (see Exercise 2.2).

It is useful to have available certain ways of combining sets to form new sets. To this end, we start with a set, $C$, whose elements are sets. Caution: there is no "set of *all* sets"!

THEOREM 2.4 *There is no set, $C$, such that, if $A$ is a set, then $A \in C$.*

PROOF: Suppose there is a set $C$ such that, if $A$ is a set, then $A \in C$. By the Axiom of Specification, $C$ has a subset

$$D = \{A \in C | A \notin A\}.$$

Then $D \in D$ if and only if $D \notin D$. Contradiction! Thus, there is no such set $C$.

The next theorem ensures the existence of a "greatest common subset" for a given set of sets.

THEOREM 2.5 *Let $C$ be a non-empty set of sets. Then there is a unique set $\Pi$ such that $x \in \Pi$ if and only if $x \in A$ for all $A \in C$.*

PROOF: Since $C$ is non-empty, there is a set $\bar{A} \in C$. By the Axiom of Specification, $\bar{A}$ has a subset

$$\Pi = \{x \in \bar{A} | x \in A \text{ for all } A \in C\}.$$

Obviously, if $x \in \Pi$, then $x \in A$ for all $A \in C$; and if $x \in A$ for all $A \in C$, then $x \in \bar{A}$, hence $x \in \Pi$. Thus $\Pi$ has the required property: $x \in \Pi$ if and only if $x \in A$ for all $A \in C$. If $\Pi'$ is another set with this property, then $\Pi' = \Pi$, by the Axiom of Identity.

DEFINITION 2.1 If $C$ is a set of sets, then the set $\Pi$ such that $x \in \Pi$ if and only if $x \in A$ for all $A \in C$ is the *intersection,* $\bigcap_{A \in C} A$, *of the sets, A, in C.*

The following theorem describes more precisely the role of $\bigcap_{A \in C} A$ as the "greatest common subset" of the sets in $C$.

THEOREM 2.6 *If C is a non-empty set of sets and X is a set, then* $X = \bigcap_{A \in C} A$ *if and only if*

1) $X \subset A$ *for all* $A \in C$ *and*
2) *if T is a set such that* $T \subset A$ *for all* $A \in C$, *then* $T \subset X$.

(Note: 1) states that $X$ is a common subset of the sets of $C$, and 2) states that every common subset, $T$, of the sets of $C$ is a subset of $X$. This is the sense in which $X$ is the "greatest" common subset.)

PROOF: Suppose $X = \bigcap_{A \in C} A$. Then $X$ satisfies 1), for: if $x \in X$, then $x \in A$ for all $A \in C$, whence $X \subset A$ for all $A \in C$. Also, $X$ satisfies 2), for: let $T$ be a set such that $T \subset A$ for all $A \in C$. If $t \in T$, then $t \in A$ for all $A \in C$, hence $t \in X$; but then $T \subset X$.

Conversely, suppose $X$ is a set satisfying 1) and 2). Let $X' = \bigcap_{A \in C} A$. Since $X'$ satisfies 1) and $X$ satisfies 2), $X' \subset X$. Since $X$ satisfies 1) and $X'$ satisfies 2), $X \subset X'$. Hence $X = X' = \bigcap_{A \in C} A$.

The following axiom will enable us to define a "least common superset" for a given set of sets.

AXIOM OF INCLUSION If $C$ is a non-empty set of sets, then there is a set $B$ such that $A \subset B$ for each $A \in C$.

If we extend our terminology to say that, if $A \subset B$, then $B$ is a "superset" of $A$, the Axiom of Inclusion can be restated thus: if $C$ is a non-empty set of sets, then there is a common superset for the sets in $C$.

Our next theorem ensures the existence of a "least common superset" for a given set of sets.

THEOREM 2.7 *If C is a set of sets, then there is a unique set* $\Sigma$ *such that* $x \in \Sigma$ *if and only if* $x \in A$ *for some* $A \in C$.

PROOF: If $C$ is a set of sets, then there is a unique set $B$ such that $x \in \Sigma$ each $A \in C$. Thus, if $x \in A$ for some $A \in C$, then $x \in B$. By the Axiom of Specification, $B$ has a subset

$$\Sigma = \{x \in B \mid x \in A \text{ for some } A \in C\}.$$

Obviously, if $x \in \Sigma$, then $x \in A$ for some $A \in C$. Conversely, if $x \in A$ for some $A \in C$, then $x \in B$, and therefore $x \in \Sigma$. Thus $\Sigma$ has the required property: $x \in \Sigma$ if and only if $x \in A$ for some $A \in C$. If $\Sigma'$ is another set with the property: $x \in \Sigma'$ if and only if $x \in A$ for some $A \in C$, then $\Sigma = \Sigma'$, by the Axiom of Identity.

DEFINITION 2.2 If $C$ is a set of sets, then the set $\Sigma$ such that $x \in \Sigma$ if and only if $x \in A$ for some $A \in C$ is called the *union*, $\bigcup_{A \in C} A$, *of the sets A in C.*

The following theorem describes more precisely the role of $\bigcup_{A \in C} A$ as the "least common superset" of the sets of $C$.

THEOREM 2.8 *If $C$ is a set of sets and $X$ is a set, then $X = \bigcup_{A \in C} A$ if and only if*

1) *$A \subset X$ for each $A \in C$*

*and*

2) *if $B$ is a set such that $A \subset B$ for each $A \in C$, then $X \subset B$.*

(Note: 1) asserts that $X$ is a common superset of the sets of $C$; 2) asserts that $X$ is a subset of every common superset of the sets in $C$.)

PROOF: Suppose $X = \bigcup_{A \in C} A$. Then $X$ satisfies 1), for: if $A \in C$ and $x \in A$, then $x \in X$; hence $A \subset X$ for each $A \in C$. Also, $X$ satisfies 2), for: let $B$ be a set such that $A \subset B$ for each $A \in C$. If $x \in X$, then $x \in A$ for some $A \in C$, hence $x \in B$, and so $X \subset B$.

Conversely, suppose $X$ is a set satisfying 1) and 2). Let $X' = \bigcup_{A \in C} A$. Since $X'$ satisfies 1) and $X$ satisfies 2), $X \subset X'$. Since $X$ satisfies 1) and $X'$ satisfies 2), $X' \subset X$. Hence $X = X' = \bigcup_{A \in C} A$.

If $A$ and $B$ are sets, we have as yet nothing to tell us if there is a set with $A$ and $B$ as elements. This is embarrassing if we wish to consider the union of $A$ and $B$. The following axiom permits us to do so.

AXIOM OF PAIRS *If $A$ and $B$ are sets, there is a set $C$ such that $A \in C$ and $B \in C$.*

Now, by the Axiom of Specification, there is a subset of $C$ consisting of just $A$ and $B$. This set $\{A, B\}$ is called a *pair.* If $A = B$, then the pair $\{A, B\} = \{A, A\} = \{A\}$ is called the *singleton* of $A$.

COROLLARY If $A$, $B$ are sets, then there is a set which is the intersection of $A$ and $B$ and there is a set which is the union of $A$ and $B$. (See Exercise 2.3).

*Notation:* We write $A \cap B$ and $A \cup B$, respectively, for the intersection and the union of $A$ and $B$.

AXIOM OF POWERS *If $A$ is a set, there is a set $\mathscr{P}(A)$ (called the power set of $A$) whose elements are the subsets of $A$.*

For example, the power set of the set whose elements are $a$, $b$, $c$ is the set whose elements are

$$\phi, \{a\}, \{b\}, \{c\}, \{a, b\}, \{b, c\}, \{c, a\}, \{a, b, c\}.$$

If $a \in A$, then one of the subsets of $A$ is the singleton $\{a\}$. We note that if $a \in A$, then $\{a\} \subset A$ and $\{a\} \in \mathscr{P}(A)$. If $a \in A$ and, $b \in B$, then the pair $\{a, b\}$ is a subset of $A \cup B$ and an element of $\mathscr{P}(A \cup B)$.

A very useful concept is that of "ordered pair." The coordinates $(x, y)$ of a point in the plane, for example, form an ordered pair. To specify a point $P$ in the plane, it is sufficient to state which two numbers will serve as the coordinates of $P$ and which one of these two numbers will be the $x$-coordinate. In the symbol (3, 2), the $x$-coordinate is *singled out* by being written first. A definition of "ordered pair" which makes use of the idea of singling out one of the elements of a pair without presupposing any notion of "first" or "second" was given by Norbert Wiener:

DEFINITION 2.3 If $a \in A$ and $b \in B$, then the *ordered pair* $(a, b)$ is the set $\{\{a\}, \{a, b\}\}$ consisting of the pair $\{a, b\}$ and the singleton $\{a\}$.*

This definition has the advantage of presupposing only the set axioms so that such concepts as order and mapping can later be defined in terms of "ordered pair" without danger of circularity.

The most important fact about Definition 2.3 is that the ordered pairs so defined behave exactly as ordered pairs should:

THEOREM 2.9 *Two ordered pairs $(a, b)$ and $(a', b')$ are equal if and only if $a = a'$, and $b = b'$.*

PROOF: If $a = a'$ and $b = b'$, then, by the Axiom of Identity,

$$(1) \quad \{\{a\}, \{a, b\}\} = \{\{a'\}, \{a', b'\}\}.$$

Conversely, suppose that (1) holds. If $a = b$, then, by Definition 2.3, and by the agreement on notation (page 3),

$$(a, b) = (a, a) = \{\{a\}, \{a, a\}\} = \{\{a\}\} = \{\{a'\}, \{a', b'\}\}.$$

*We note that if $a = b$, then $(a, b) = (a, a) = \{\{a\}, \{a, a\}\} = \{\{a\}, \{a\}\} = \{\{a\}\}$. This result may seem strange, but it does no harm.

By the Axiom of Identity, $a = a' = b'$. Since $a = b$, $a = a'$ and $b = b'$. If $a \neq b$, then, by the Axiom of Identity, $\{a, b\} \neq \{a'\}$. Hence

$$\{a, b\} = \{a', b'\}, \text{ and } \{a\} = \{a'\}.$$

But then $a = a'$ and $b = b'$.

THEOREM 2.10 *If $A$ and $B$ are non-empty sets, then there is a set $C$ consisting of all ordered pairs $(a, b)$ with $a \in A$ and $b \in B$.*

PROOF: Each ordered pair $(a, b) = \{\{a\}, \{a, b\}\}$ is a subset of the power set $\mathscr{P}(A \cup B)$. By the Axiom of Specification, the condition "$x$ is an ordered pair $(a, b)$ with $a \in A$ and $b \in B$" determines a subset $C$ of the power set $\mathscr{P}(\mathscr{P}(A \cup B))$ of $\mathscr{P}(A \cup B)$. The set $C$ consists of all ordered pairs $(a, b)$ with $a \in A$ and $b \in B$.

DEFINITION 2.4 The set $C$ of Theorem 2.10 is called the *Cartesian product* of $A$ and $B$. We denote it by "$A \times B$."

*Example 1:* If $A$ and $B$ are both the set of all real numbers, then $A \times B$ is the Cartesian plane. (The term "Cartesian product" is taken from this example.)

*Example 2:* If $A$ is the set of all real numbers and $B$ is the set of all integers, then $A \times B$ is the subset of the Cartesian plane consisting of all points lying on the lines $y = n$ where $n$ is any integer.

*Example 3:* If $A$ is the set of all positive integers and $B$ is the set of all integers, then $A \times B$ is the set of all lattice points in the right half-plane.

*Exercise 2.1* Prove Theorem 2.1.

*Exercise 2.2* If $A$ is any set, then $\phi \subset A$.

*Exercise 2.3* If $A$, $B$ are sets, then there are sets $\Pi$ and $\Sigma$ such that $\Pi = A \cap B$ and $\Sigma = A \cup B$.

*Exercise 2.4*

(a) If $A$, $B$ are sets, then $A \cap B = B \cap A$ and $A \cup B = B \cup A$.

(b) If $A$, $B$, $C$ are sets, then

$$A \cap (B \cap C) = (A \cap B) \cap C$$
$$A \cup (B \cup C) = (A \cup B) \cup C.$$

(c) If $A$, $B$, $C$ are sets, then

$$A \cap (B \cup C) = (A \cap B) \cup (A \cap C)$$
$$A \cup (B \cap C) = (A \cup B) \cap (A \cup C).$$

*Exercise 2.5* If $A$, $B$ are sets, then the following statements are equivalent:

(1) $A \subset B$

(2) $A \cup B = B$

(3) $A \cap B = A$.

*Exercise 2.6* If $A$ is a set, then the following statements are equivalent:

(1) $A = \phi$.

(2) There is a set $C$ such that $A \cup B = B$ for all $B \subset C$.

(3) There is a set $C$ such that $A \cap B = A$ for all $B \subset C$.

*Exercise 2.7* If $A$, $B$ are sets, and $A - B = \{x \in A \,|\, x \notin B\}$, then

(1) $A - B = A - (A \cap B)$,

(2) $A - B = \phi$ if and only if $A \subset B$.

*Exercise 2.8*

(a) If $A$, $B$, $C$ are non-empty sets, then

$$A \times (B \cup C) = (A \times B) \cup (A \times C).$$

(b) There are sets $A$, $B$, $C$ such that

$$A \times (B \times C) \neq (A \times B) \times C.$$

Hint: use the fact that $\phi \neq \{\phi\}$.

(c) $A \times B = B \times A$ if and only if $A = B$.

In fact, the following statements are equivalent:

$$A \times B \subset B \times A$$
$$B \times A \subset A \times B$$
$$A \times B = B \times A$$
$$A = B.$$

3. **BINARY RELATIONS.** If $A$ and $B$ are sets, then a condition $Q(x)$ determines, by the Axiom of Specification, a subset $R$ of $A \times B$ consisting of all ordered pairs $x = (a, b)$ for which the condition holds. Such a condition expresses what, in ordinary language, is called a relation between $a$ and $b$. Examples are "$a$ is less than $b$," "$a$ equals $b$," "$a$ is divorced from $b$." In mathematics it is convenient to identify the set $R$ with the relation. In fact, if $R$ is any subset of $A \times B$, one may think of the condition $Q(x)$: "$x \in R$" as expressing a relation between $a$ and $b$, where $x = (a, b)$.

DEFINITION 3.1 A *binary relation* is a subset $R$ of a Cartesian product $A \times B$. If $R \subset A \times A$, we call $R$ a binary relation *in* $A$. If $(a, b) \in R$, we may write "$a \, R \, b$."

DEFINITION 3.2 Let $R$ be a binary relation defined in a set $A$. Then,

(a) $R$ is *reflexive* if $aRa$ holds for all $a \in A$.

(b) $R$ is *symmetric* if $bRa$ holds whenever $aRb$ holds for $a, b \in A$.
(c) $R$ is *transitive* if $aRc$ holds whenever $aRb$ and $bRc$ both hold for $a, b, c \in A$.
(d) $R$ is *anti-symmetric* if $a = b$ whenever $aRb$ and $bRa$ both hold for $a, b \in A$.
(e) $R$ satisfies the *law of trichotomy* if, for any $a, b \in A$, exactly one of $aRb$, $bRa$, and $a = b$ holds.

DEFINITION 3.3 A binary relation $R$ in a set $A$ is an *equivalence relation in A* if $R$ is reflexive, symmetric and transitive. If $R$ is an equivalence relation in $A$, then for $a \in A$, the set

$$C_a = \{x \in A \mid x \, R \, a\}$$

is the *equivalence class of a, with respect to R.* The set of all equivalence classes with respect to $R$ is *the factor set, A/R, of A modulo R.*

THEOREM 3.1 *If R is an equivalence relation in A and a, b ∈ A, then* $C_a = C_b$ *if and only if* $aRb$.

PROOF: Suppose $C_a = C_b$. Since $R$ is reflexive, $aRa$ and $a \in C_a = C_b$. Hence $aRb$.

Suppose $aRb$. If $x \in C_a$, then $xRa$. Since $R$ is transitive, it follows that $xRb$, and so $x \in C_b$. Thus, $C_a \subset C_b$. Since $R$ is symmetric, we have $bRa$. Interchanging $a$ and $b$ in the above argument, we obtain $C_b \subset C_a$. Hence $C_a = C_b$.

The most familiar example of an equivalence relation is equality of numbers. Other examples from elementary mathematics include similarity and congruence of triangles. As a further illustration, we give the following example from arithmetic: let $A$ be the set of all integers. Define a binary relation $R$ in $A$ by: $aRb$ if and only if 3 is a divisor of $a - b$. It is easily checked that $R$ is an equivalence relation in $A$ (see Exercise 3.1). If $aRb$, we say that $a$ is congruent to $b$ modulo 3 and write $a \equiv b \bmod 3$. The equivalence classes with respect to congruence modulo 3 are $C_0 = \{x \in A \mid x \equiv 0 \bmod 3\}$, $C_1 = \{x \in A \mid x \equiv 1 \bmod 3\}$ and $C_2 = \{x \in A \mid x \equiv 2 \bmod 3\}$. Since every integer is either a multiple of 3, 1 plus a multiple of 3, or 2 plus a multiple of 3, we have $A = C_0 \cup C_1 \cup C_2$; and, since no two of the above conditions can hold simultaneously, the classes $C_0$, $C_1$ and $C_2$ are pairwise disjoint. This illustrates the fact that the equivalence classes with respect to an equivalence relation $R$ in a set $A$ form a partition of the set $A$, in the sense of the following definition.

DEFINITION 3.4 A set $P$ of non-empty subsets of a set $A$ is a *partition of A* if

1) $A = \bigcup_{X \in P} X$

and

2) $X \cap Y = \phi$ for $X \neq Y$ $(X, Y \in P)$.

THEOREM 3.2 *Fundamental Partition Theorem*

1) *If $R$ is an equivalence relation in a set $A$, then the equivalence classes with respect to $R$ form a partition of $A$.*

2) *Every partition, $P$, of a set $A$ is the set of all equivalence classes with respect to some equivalence relation $R$ in $A$.*

PROOF: 1) If $a \in A$, then, since $R$ is reflexive, $a \in C_a$. Hence $A \subset \bigcup_{a \in A} C_a$. But, clearly, $\bigcup_{a \in A} C_a \subset A$, and so $A = \bigcup_{a \in A} C_a$. For $a, b \in A$, suppose $C_a \cap C_b \neq \phi$. If $c \in C_a \cap C_b$, then $c\ R\ a$ and $c\ R\ b$. By Theorem 3.1, it follows that $C_a = C_c = C_b$. Thus, the set $P = \{C_a | a \in A\}$ is a partition of $A$.

2) Suppose $P$ is a partition of $A$. Define a binary relation $R$ in $A$ by: $a\ R\ b$ if $a, b \in X$ for some $X \in P$. Obviously, $R$ is reflexive and symmetric. We show that $R$ is transitive. Suppose $a\ R\ b$ and $b\ R\ c$ hold for $a, b, c \in A$. Then there are $X, Y \in P$ such that $a, b \in X$ and $b, c \in Y$. Since $c \in X \cap Y$, $X \cap Y \neq \phi$. Hence $X = Y$, and $a, c \in X$. But then $a\ R\ c$, and $R$ is transitive. Thus, $R$ is an equivalence relation in $A$.

The factor set, $A/R$, of $A$ modulo $R$ is equal to $P$. For: if $a \in A$, then $a \in X$ for some $X \in P$, hence $C_a = \{x \in A | x\ R\ a\} = \{x \in A | x, a \in X\} = X$. Hence $A/R \subset P$. For each $X \in P$, since $X \neq \phi$, there is some $a \in X$ and, by the preceding result, $X = C_a$. But then $P \subset A/R$, and the equality follows.

DEFINITION 3.5 If a binary relation in a set $A$ is reflexive, antisymmetric, and transitive, it is called a *partial order relation.*

The relation "$\leqq$" for real numbers and the relation "is a divisor of" for positive integers are examples of partial order relations.

Finally, "$<$" for real numbers is an example of a third kind of relation.

DEFINITION 3.6 A binary relation $R$ defined in a set $A$ is called an *order relation* in $A$ if it is transitive and satisfies the law of trichotomy. A set in which is defined an order relation is called an *ordered set.*

*Exercise 3.1*

(a) Prove that the binary relation "$\equiv$ mod 3" defined on the set, $A$, of all integers by: $a \equiv b$ mod 3 if $a - b$ is a multiple of 3 is an equivalence relation in $A$.

(b) Identify $C_{2864}$ as one of the classes $C_0, C_1$ or $C_2$.

*Exercise 3.2* What is wrong with the following "proof"?

"If $R$ is a binary relation in $A$ which is symmetric and transitive, then it is also reflexive, for: by symmetry, $a\ R\ b$ implies $b\ R\ a$; hence, by transitivity, $a\,R\,a$."

Give an example of a binary relation which is symmetric and transitive, but not reflexive.

*Exercise 3.3* Let $\mathscr{P}(A)$ be the power set of a set $A$ and let

$$R = \{(A, B) \in \mathscr{P}(A) \times \mathscr{P}(A) \,|\, A \subset B\}.$$

Prove that $R$ is a partial order relation in $\mathscr{P}(A)$.

*Exercise 3.4*

(a) Let $A$ be a set, $\sim$ an equivalence relation in $A$ and $<$ an order relation in $A$. If, for $x, y \in A$, "$x \lesssim y$" is defined to mean $x < y$ or $x \sim y$, prove that $\lesssim$ is a partial order relation in $A$.

(b) Derive the relation, $\subset$, of set inclusion (see the preceding exercise) from an order relation and an equivalence realtion in the manner described in (a).

## 4. FUNCTIONS

DEFINITION 4.1 Let $X$, $Y$ be non-empty sets and let $F$ be a subset of $X \times Y$. Then $F$ is a *function from $X$ to $Y$* ($F: X \to Y$) if, for each $x \in X$, there is exactly one $y \in Y$ such that $(x, y) \in F$.

Let $F: X \to Y$. Then $X$ is the *domain* of $F$ and $Y$ is the *co-domain* of $F$. If $(x, y) \in F$, we write $y = F(x)$. The set $F(X) = \{F(x) \,|\, x \in X\}$ is the *range* (or *image*) of $F$. $F$ is an *injection* (or *$F$ is injective,* or *$F$ is 1–1*) if, for $x, x' \in X$, $F(x) = F(x')$ implies $x = x'$. $F$ is a *surjection* (or *$F$ is surjective,* or *$F$ maps $X$ onto $Y$*) if $F(X) = Y$. A function $F: X \to Y$ which is both an injection and surjection is a *bijection.*

If $F$ and $G$ are functions, then $F$ and $G$ are *the same function* if they have the same domain and the same co-domain, and if $F$ and $G$ are the same set of ordered pairs. (Thus, if $F: X \to Y$ and $G: S \to T$ are functions, then $F$ and $G$ are the same function if and only if $X = S$, $Y = T$ and $F(a) = G(a)$ for each $a \in A$.)

Functions are also referred to as *mappings*, or *maps*, and in some contexts as *transformations.*

*Example 1:* Let $X$ be the set of all students enrolled in a given course and let $Y$ be the alphabet. Definé $G: X \to Y$ by: $G(x)$ is the letter grade received by student $x$ at the end of the term. Then $G$ is a function from $X$ to $Y$. In general, this function is neither injective nor surjective. (Why?)

*Example 2:* Let $X$ be the set of all real numbers, and let $Y$ be the set of all non-negative real numbers. Then the function $F: X \to X$ defined by $F(x) = x^2$

for each $x \in X$ is neither surjective nor injective; the function $G\colon Y \to X$ defined by $G(x) = x^2$ for each $x \in Y$ is injective, but not surjective; the function $H\colon X \to Y$ defined by $H(x) = x^2$ for each $x \in X$ is surjective, but not injective; and the function $K\colon Y \to Y$ defined by $K(x) = x^2$ for each $x \in Y$ is both injective and surjective, i.e., $K$ is a bijection.

Note that no two of $F$, $G$, $H$ and $K$ are the same function, even though, as sets of ordered pairs, $F = H$ and $G = K$.

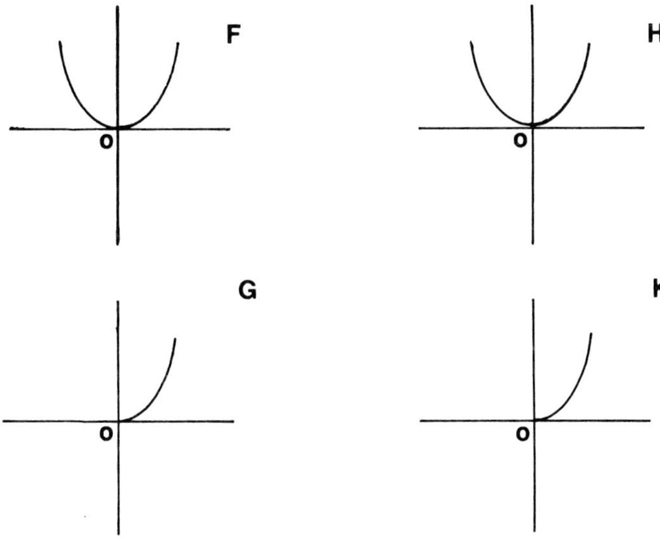


*Example 3:* If $X$ is any set, then there is a function $F\colon X \to X$ defined by: $F(x) = x$ for each $x \in X$. This function is obviously a bijection. For instance, if $X$ is the set of all real numbers, then the graph of $F$ is:

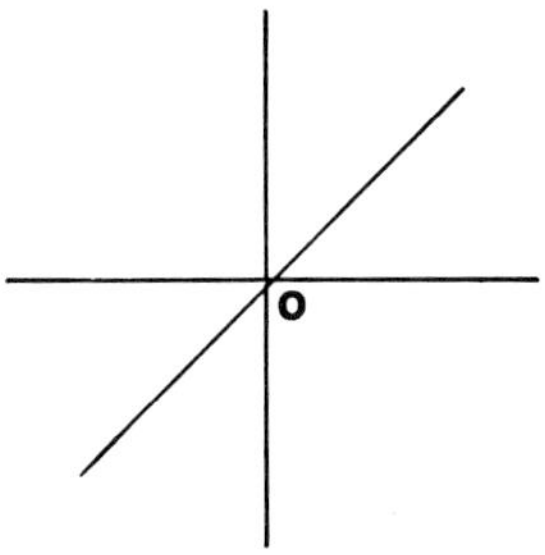

DEFINITION 4.2 Let $X$ be any set. Then the function $F: X \to X$ defined by $F(x) = x$ for each $x \in X$ is the *identity function*, $I_X$, *on* $X$.

DEFINITION 4.3 Let $X$, $Y$ be sets and let $X'$, $Y'$ be subsets, respectively, of $X$ and $Y$. Let $F: X \to Y$ be a function. Then the function $G: X' \to Y$ defined by: $G(x) = F(x)$ for each $x \in X'$ is the *restriction*, $F|_{X'}$, *of* $F$ *to* $X'$. If $F(X) \subset Y'$, then the function $H: X \to Y'$ defined by: $H(x) = F(x)$ for each $x \in X$ is the *co-restriction*, ${}_{Y'}|F$, *of* $F$ *to* $Y'$.

(For instance, in Example 2, above, $G$ is the restriction of $F$ to $Y$, $H$ is the co-restriction of $F$ to $Y$, $K$ is the restriction of $H$ to $Y$ and also the co-restriction of $G$ to $Y$.)

Now let $F: X \to Y$ and $G: Y \to Z$. Then we may define a function $H: X \to Z$ by $H(x) = G(F(x))$ for each $x \in X$. It is clear that $H$ *is* a function from $X$ to $Z$ since the set $H = \{(x, G(F(x)) | x \in X\}$ has the property: for each $x \in X$, there is exactly one $z \in Z$ such that $(x, z) \in H$.

DEFINITION 4.4 If $F: X \to Y$ and $G: Y \to Z$, then the function $H: X \to Z$ defined by $H(x) = G(F(x)$ for) each $x \in X$ is the *composite*, $GF$, of $F$ and $G$.

If $F: X \to Y$, $G: Y \to Z$ and $H: X \to Z$ is the composite of $F$ and $G$, we may portray this fact by a diagram:

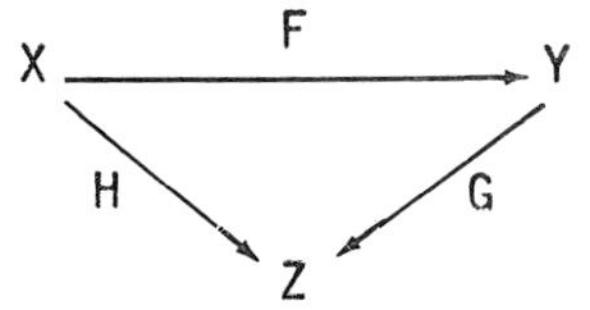


Since $H = GF$, we may take two alternate routes in sending elements of $X$ to elements of $Z$:

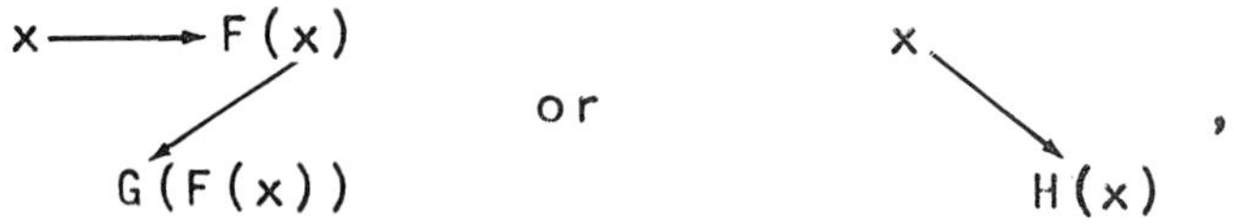


both times achieving the same effect. For this reason, we say that the diagram, above, *commutes*.

THEOREM 4.1 *Composition of functions is "associative," i.e., if* $F: X \to Y$, $G: Y \to Z$, *and* $H: Z \to T$, *then* $(HG)F = H(GF)$.

PROOF: First note that the functions $(HG)F$ and $H(GF)$ have the same domain, $X$, and the same co-domain, $T$. Also, for each $x \in X$, $((HG)F)(x)$

$= (HG)(F(x)) = H(G(F(x)))$, and $(H(GF))(x) = H((GF)(x)) = H(G(F(x)))$. But then $(HG)F$ and $H(GF)$ are the same function.

DEFINITION 4.5 Let $X$, $Y$ be sets, $F: X \to Y$ and $G: Y \to X$. Then
$G$ is a *right inverse function for F* if $FG = I_Y$
(i.e., if $F(G(y)) = y$ for each $y \in Y$);
$G$ is a *left inverse function for F* if $GF = I_X$
(i.e., if $G(F(x)) = x$ for each $x \in X$);
$G$ is a *(2-sided) inverse function for F* if $FG = I_Y$ and $GF = I_X$.

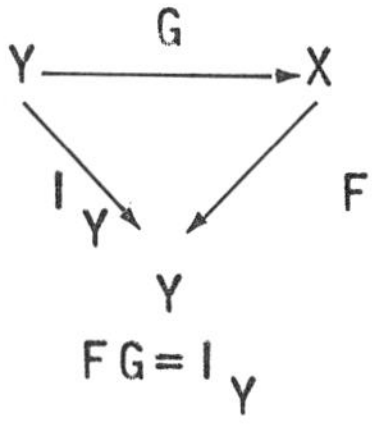


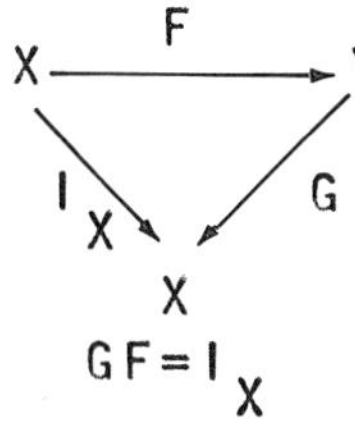


COROLLARY A function $F: X \to Y$ has at most one 2-sided inverse function.

PROOF: If $G_1: Y \to X$ and $G_2: Y \to X$ are both 2-sided inverse functions for $F$, then $G_1 = G_1 I_Y = G_1(FG_2) = (G_1 F)G_2 = I_X G_2 = G_2$.

*Notation:* If $F: X \to Y$ has a 2-sided inverse function $G: Y \to X$, we write $F^{-1} = G$.

One-sided inverse functions are not unique. For example, if $X$ is the set of all real numbers, $Y$ the set of all non-negative real numbers, and $H: X \to Y$ the function defined by $H(x) = x^2$ for each $x \in X$, then two distinct right inverse functions for $H$ are the functions $G_1: Y \to X$ and $G_2: Y \to X$ defined by $G_1(y) = \sqrt{y}$ for each $y \in Y$, and $G_2(y) = -\sqrt{y}$ for each $y \in Y$. (Note that, for each $y \in Y$, $(FG_1)(y) = (\sqrt{y})^2 = y$ and $(FG_2)(y) = (-\sqrt{y})^2 = y$, hence $FG_1 = I_Y = FG_2$.

THEOREM 4.2 *A function $F: X \to Y$ has a 2-sided inverse function if and only if it is a bijection.*

PROOF: a) Suppose $F: X \to Y$ has a 2-sided inverse function $G: Y \to X$. Then, for $x, x' \in X$, $F(x) = F(x')$ implies $G(F(x)) = G(F(x'))$, hence $GF(x) = GF(x')$, and $x = x'$. Thus, $F$ is injective.

For each $y \in Y$, $y = I_Y(y) = (GF)(y) = F(G(y))$. Thus, $F$ is surjective. It follows that $F$ is a bijection.

b) Conversely, suppose $F: X \to Y$ is a bijection. Define $G: Y \to X$ by $G(y) = x$ if $F(x) = y$. Then, since $F$ is both surjective and injective, $G$ is a function with domain $Y$ and co-domain $X$. For each $x \in X$, $G(F(x)) = G(y)$

$= x$; for each $y \in Y$, $F(G(y)) = F(x) = y$. Thus, $GF = I_X$ and $FG = I_Y$. Hence $G$ is a 2-sided inverse function for $F$.

If we examine part a) of the proof above, we observe that we have, in fact, established the results: 1) if $F: X \to Y$ has a left inverse function, then $F$ is injective; and 2) if $F: X \to Y$ has a right inverse function, then $F$ is surjective. Are the converses of 1) and 2) true?

Let us first examine the converse of 1): Suppose $F: X \to Y$. Since $F$ is injective, there is, for each $y \in F(X)$, a unique element $x \in X$ such that $y = F(x)$. As in the proof of Theorem 4.2, we may define $G(y) = x$ if $F(x) = y$. We must find a way to complete the definition of a function $G$ from $Y$ to $X$ in case $F$ is not surjective, i.e., in case $F(X) \neq Y$. If we are given a fixed element $x_0 \in X$, we may define $G(y) = x_0$ for each $y \in Y - F(X)$. Then $G$ is a function from $Y$ to $X$ and $GF(x) = G(F(x)) = x$ for each $x \in X$, whence $GF = I_X$ and $G$ is a left inverse function of $F$, as desired. Observe, however, that this definition of $G$ was accomplished by assuming the possibility of selecting a specific element, $x_0$, from $X$.

Let us, next, examine the converse of 2). Here, $F: X \to Y$ is surjective, hence for each $y \subset Y$, there is a non-empty set $X_y \in X$, where $X_y = \{x \in X | F(x) = y\}$. If we are given an element $x_y \in X_y$, for each $y \in Y$, then we can define $G: Y \to X$ by $G(y) = x_y$ for each $y \in Y$. Clearly, $F(G(y)) = F(x_y) = y$ for each $y \in Y$, and so $FG = I_Y$, whence $G$ is a right inverse function of $F$. But note that this definition of $G$ depends on the possibility of *choosing a specific element* from each of the sets $X_y$. The set theory we have developed thus far provides us with no means for accomplishing this selection. The following additional axiom, due to Zermelo (1904), removes these difficulties.

AXIOM OF CHOICE Let $A$ be a non-empty set. Then there is a function $T: \mathscr{P}(A) - \{\phi\} \to A$ such that $T(X) \in X$ for each non-empty subset $X$ of $A$.

Note that $T$ associates with each non-empty subset, $X$, of $A$ exactly one element $T(X) \in X$. We call $T$ a *choice function* for $A$. Throughout this book, we shall find numerous uses for the Axiom of Choice. While this axiom may seem intuitively obvious, it cannot be proved from the other set theoretic axioms we have assumed. This independence of the Axiom of Choice from basic set theory was proved only recently (1962) by Paul Cohen.

Our first application of the Axiom of Choice is the theorem we discussed above.

THEOREM 4.3 *If $F: X \to Y$, then $F$ is injective if and only if $F$ has a left inverse function and $F$ is surjective if and only if $F$ has a right inverse function. (See Exercise 4.7).*

*Exercise 4.1*

(1) Prove that the composite of two injections is an injection.
(2) Prove that the composite of two surjections is a surjection.
(3) Conclude that the composite of two bijections is a bijection.

*Exercise 4.2* Let $X$, $Y$ be sets, $F: X \to Y$ a bijection. Let $S: X \to X$ be a function. Prove:

(1) $FSF^{-1}$ is a function from $Y$ to $Y$.
(2) $FSF^{-1}$ is an injection if and only if $S$ is an injection.
(3) $FSF^{-1}$ is a surjection if and only if $S$ is a surjection.
(4) $FSF^{-1}$ is a bijection if and only if $S$ is a bijection.

*Exercise 4.3* Let $\mathscr{P}(X)$ be the power set of a set $X$. Or $\mathscr{P}(X) - \{\phi\}$, define $R$ by: $ARB$ if there is a bijection $F: A \to B$. Prove that $R$ is an equivalence relation in $\mathscr{P}(X) - \{\phi\}$.

*Exercise 4.4* Let $F: X \to Y$ and $G: Y \to Z$.

(1) If $GF$ is injective, need $F$ or $G$ be injective?
(2) If $GF$ is surjective, need $F$ or $G$ be surjective?

*Exercise 4.5*

(a) Let $F$, $G_1$, $G_2$ be functions such that $FG_1 = FG_2$. Prove: if $F$ is injective, then $G_1 = G_2$.
(b) Let $F$, $G_1$, $G_2$ be functions such that $G_1F = G_2F$. Prove: if $F$ is surjective, then $G_1 = G_2$.

*Exercise 4.6* Let $F: X \to Y$. On $X$, define a binary relation $R$ by: $x\,R\,x'$ if $F(x) = F(x')$.

(a) Prove that $R$ is an equivalence relation in $X$.
(b) For each $y \in F(X)$, let $X_y = \{x \in X \mid F(x) = y\}$.
Let $Q = \{X_y \mid y \in F(X)\}$. Prove that $X/R = Q$.
(c) Define $G: F(X) \to Q$ by: $G(y) = X_y$ for each $y \in F(X)$. Prove that $G$ is a bijection.
(d) Let $T$ be a choice function for $X$. Prove: if $F$ is surjective, then $TG$ is a right inverse function for $F$.

*Exercise 4.7* Prove Theorem 4.3, making explicit use of a choice function.

# CHAPTER 1

# THE NATURAL NUMBERS

**1. THE NATURAL NUMBERS.** Familiar mathematics begins with the whole numbers. We give here a system of axioms from which the familiar properties of the whole numbers can be proved as theorems. For guidance in our choice of axioms, we examine the list of whole numbers

$$1, 2, 3, 4, \ldots .$$

We observe that every whole number has one, and only one, successor in the list, and that every whole number except 1 is the successor of one, and only one, whole number. We observe, too, that we can obtain all whole numbers by starting with 1 and taking successive successors. These properties will be reflected in the following axiom.

AXIOM 1.1 There is a set $\mathbf{N}$ and there is a function $S\colon \mathbf{N} \to \mathbf{N}$ such that

A1) $S$ is injective;

A2) $S$ is not surjective;

A3) If $u \in \mathbf{N} - S(\mathbf{N})$, and if $M \subset \mathbf{N}$ such that

$$u \in M$$

and

$$n \in M \Rightarrow S(n) \in M$$

then $M = \mathbf{N}$.

(We refer to the elements of $\mathbf{N}$ as *natural numbers* and to $S$ as the *successor function*.)

In the following, we show that there is just one natural number outside the range of $S$. This natural number will play the role of the whole number 1. In the presence of our axioms, it will be possible to introduce exactly one binary operation with the properties of familiar addition, and exactly one binary operation with the properties of familiar multiplication. Besides these two operations, we also introduce in $\mathbf{N}$ an order relation corresponding to the familiar order for whole numbers. Finally, we use the natural numbers in formulating a definition of "finite set," and show how the natural numbers "count" finite sets in a way which corresponds to the familiar process of counting with whole numbers.

THEOREM 1.1 *If* $u \in \mathbf{N} - S(\mathbf{N})$, then $\mathbf{N} = \{u\} \cup S(\mathbf{N})$.

PROOF: Let $M = \{u\} \cup S(\mathbf{N})$. Then $u \in M$ and, for each $n \in M$, $S(n) \in M$. By A3 of Axiom 1.1, $M = \mathbf{N}$. Thus, $\mathbf{N} = \{u\} \cup S(\mathbf{N})$.

COROLLARY There is exactly one non-successor in **N**.

*Notation:* We denote the non-successor by the symbol "1".

Axiom 1.1, A3, can now be restated as follows:
A3′: *Principle of Induction* If $M \subset \mathbf{N}$ such that

$$1 \in M$$

and

$$n \in M \Rightarrow S(n) \in M$$

then $M = \mathbf{N}$.

A subset $M$ of **N** with the property: $n \in M \Rightarrow S(n) \in M$ is called an *inductive set.* In this terminology, the Principle of Induction states that any inductive set containing 1 is the set of all natural numbers.

*Exercise 1.1* Prove that, for each $n \in \mathbf{N}$, $S(n) \neq n$.

*Exercise 1.2* Prove that conditions A1, A2 and A3 of Axiom 1.1 are independent, i.e., for each $i \neq j$ $(i, j = 1, 2, 3)$ there is a set $\mathbf{N}_{ij}$ and there is a function $S_{ij}$: $\mathbf{N}_{ij} \to \mathbf{N}_{ij}$ satisfying conditions $A_i$ and $A_j$, but not satisfying the remaining condition.

*Exercise 1.3* Assume the axioms of Chapter 0 (except for the Axiom of Choice), together with the following axiom:

There is a set $X$ such that

1) $\phi \in X$;
2) if $A \in X$, then $A \cup \{A\} \in X$;
3) if $A, B \in X$ such that $A \in B$ and $B \in A$, then $A = B$.

Prove Axiom 1.1, i.e., prove: there is a set **N** and there is a function $S$: $\mathbf{N} \to \mathbf{N}$ satisfying A1, A2 and A3 of Axiom 1.1.

(Hint: Let **N** be the intersection of all subsets $T$ of $X$ such that

1) $\phi \in T$

and

2) if $A \in T$, then $A \cup \{A\} \in T$.

Define $S$: $\mathbf{N} \to \mathbf{N}$ by: $S(A) = A \cup \{A\}$ for each $A \in \mathbf{N}$.)

**2. RECURSION.** One of the most useful applications of the Principle of Induction is the "recursive definition" of functions with domain **N**.

DEFINITION 2.1 For any set $A$, a function $F$: $\mathbf{N} \to A$ is a *sequence in A*.

(Subscript notation is frequently used to replace functional notation in designating the elements of the range of a sequence $F$, i.e., one writes "$a_1$" for $F(1)$, "$a_2$" for $F(2)$, etc.)

A function $F$ with domain **N** and range in a set $A$ may be defined by giving an explicit condition which an ordered pair $(n, a) \in \mathbf{N} \times A$ must satisfy so that the element $a$ will be $F(n)$. For example, a mapping $F$ with range in **N** may be defined by the condition $F(n) = S(S(n))$. In many cases, however, it it impractical to specify such a condition explicitly, and it is desirable to define the function "recursively." This is done by specifying two things: (i) how to obtain $F(1)$, and (ii) how to obtain $F(S(n))$ from $F(n)$. Conditions under which a function so defined exists are given in the following theorem.

THEOREM 2.1 *(Recursion Theorem) If $G$: $A \to A$ and $a \in A$, then there is a unique function $F$: $\mathbf{N} \to A$ such that*

$$F(1) = a \text{ and } F(S(n)) = G(F(n)) \text{ for all } n \in \mathbf{N}.$$

PROOF: Let C be the set of all subsets, $H$, of $\mathbf{N} \times A$ such that

(1) $(1, a) \in H$ and if $(n, b) \in H$ then $(S(n), G(b)) \in H$.

Let

$$F = \bigcup_{H \in C} H$$

Then $F \subset \mathbf{N} \times A$ and $F$ satisfies (1). If $F'$ satisfies (1) and $F' \subset F$, then $F' = F$. This "minimal" property of $F$ is used to show that

(2) for each $n \in \mathbf{N}$ there is exactly one $b \in A$ such that $(n, b) \in F$,

from which it follows that $F$ is a function from **N** to $A$.

Let

(3) $M = \{n \in \mathbf{N} \mid (n, b) \in F$ for exactly one $b \in A\}$.

Assume that $1 \notin M$. Since $(1, a) \in F$, there is some $b \neq a$ in $A$ such that $(1, b) \in F$. Let

(4) $F_b = F - \{(1, b)\}$.

Then $F_b \subset F$ and, since $(1, b) \neq (1, a)$, $(1, a) \in F_b$. Suppose $(m, c) \in F_b$. Then $(m, c)$, $(S(m), G(c)) \in F$ and, since $(S(m), G(c)) \neq (1, b)$, $(S(m), G(c)) \in F_b$. Hence $F_b$ satisfies (1), $F_b \subset F$ and $F_b = F$, contrary to (4). Therefore

$$(5) \qquad 1 \in M.$$

Assume that, for some $n \in \mathbf{N}$, $n \in M$ and $S(n) \notin M$. By (3) there is exactly one $b \in A$ such that $(n, b) \in F$. Now, by (1), $(S(n), G(b)) \in F$ and, since $S(n) \notin M$, there is some $c \in A$ such that

$$(6) \qquad c \neq G(b) \text{ and } (S(n), c) \in F.$$

Let

$$F_c = F - \{(S(n), c)\}.$$

Then $F_c \subset F$ and, since $(S(n), c) \neq (1, a)$, $(1, a) \in F_c$. Suppose $(m, d) \in F_c$. Then $(m, d) \in F$ and, by (1), $(S(m), G(d)) \in F$. If $m = n$ then, since $(n, b) \in F$ and $n \in M$, $d = b$. Now, by (6), $(S(m), G(d)) = (S(n), G(b)) \neq (S(n), c)$ and, by (7), $(S(m), G(d)) \in F_c$. Therefore $F_c$ satisfies (1), $F_c \subset F$ and so $F_c = F$, contrary to (7). Hence

$$(8) \qquad \text{for each } n \in \mathbf{N}, \text{ if } n \in M \text{ then } S(n) \in M.$$

By (5), (8), $M = \mathbf{N}$ and (2) is proved. It follows that $F$ is a function from $\mathbf{N}$ to $A$. Since $F$ satisfies (1), $F$ has the required property.

Suppose $F' = \mathbf{N} \to A$ and $F'$ has the required property. Then $F'(1) = a = F(1)$ and if $F'(n) = F(n)$, for $n \in \mathbf{N}$, then

$$F'(S(n)) = G(F'(n)) = G(F(n)) = F(S(n)).$$

By the Principle of Induction, $F'(n) = F(n)$ for all $n \in \mathbf{N}$, hence $F = F'$. Thus $F$ is the only function from $\mathbf{N}$ to $A$ with the required property.

*Example:* We use the Recursion Theorem to define powers (iterates) of functions. Let $X$, $Y$ be sets, $H$: $X \to Y$. Then $H^m$ may be defined for each $m \in \mathbf{N}$ as follows:

$$H^1 = H,$$

$$H^{S(n)} = HH^n \text{ for each } n \in \mathbf{N}.$$

This is an example of a "recursive definition." We interpret it as an application of the Recursion Theorem:

Let F be the set of all functions from $X$ to $Y$. Then $H \in$ F. Define $G$: F $\to$ F by: $G(T) = HT$ for each $T \in$ F. By the Recursion Theorem, there is a unique function $F$: $\mathbf{N} \to$ F such that $F(1) = H$ and $F(S(n)) = G(F(n))$ for each $n \in \mathbf{N}$. Writing $H^n$ for $F(n)$, we have

$$H^1 = F(1) = H$$

and

$$H^{S(n)} = F(S(n)) = G(F(n)) = G(H^n) = HH^n$$

for each $n \in \mathbf{N}$.

The following more general theorem is useful in some cases.

*Generalized Recursion Theorem.* Let $A$ be a non-empty set, and let $a$ be an element of $A$. For each $n \in N$, let $G_n$: $A \to A$. Then there is exactly one function $F$: $N \to A$ such that

$$(1) \quad F(1) = a$$

and

$$(2) \quad F(S(n)) = G_n(F(n))$$

for all $n \in \mathbf{N}$.

The proof of this theorem may be obtained by a slight modification of the proof of the Recursion Theorem and should be an instructive exercise for the reader (Exercise 2.4). Note that the Recursion Theorem deals with the special case where the functions $G_n$ are all equal.

*Exercise 2.1* Prove that, for each $n \in \mathbf{N}$, $S(n) = S^n(1)$. (Thus, in familiar notation, $2 = S(1)$, $3 = S^2(1) = S(2)$, $4 = S^3(1) = S(3)$, etc.)

*Exercise 2.2* Let $X$ be a set, $x_0 \in X$ and $T$: $X \to X$ an injection such that $X = \{x_0\} \cup \{T^n x_0 \mid n \in \mathbf{N}\}$ and $x_0 \notin \{T^n x_0 \mid n \in \mathbf{N}\}$.

(a) Prove that there is a unique bijection $H$: $\mathbf{N} \to X$ such that $H(1) = x_0$ and $H(S(n)) = T(H(n))$ for each $n \in \mathbf{N}$.

(b) If $A$ is a set, $a \in A$, and $G$: $A \to A$, prove that there is a function $F$: $X \to A$ such that $F(x_0) = a$ and $F(T(n)) = G(F(n))$ for each $n \in \mathbf{N}$.

*Exercise 2.3* Let $A$ be a set, $a \in A$, and let $G$: $A \to A$ be a surjection. Prove that, for each $m \in \mathbf{N}$, there is a function $F_m$: $\mathbf{N} \to A$ such that $F_m(m) = a$ and $F_m(S(n)) = G(F_m(n))$ for each $n \in \mathbf{N}$. If $G$ is a bijection, prove that $F_m$ is uniquely determined.

*Exercise 2.4* (Generalized Recursion Theorem). Let $A$ be a set, $a \in A$. For each $n \in \mathbf{N}$, let $G_n$: $A \to A$ be a function. Prove: there is a unique function $F$: $\mathbf{N} \to A$ such that $F(1) = a$ and $F(S(n)) = G_n(F(n))$ for each $n \in \mathbf{N}$.

(Hint: let $F = \{T \subset \mathbf{N} \times A \mid (1, a) \in F$ and $(n, b \in F \Rightarrow (S(n), G_n(b)) \in F\}$.

## 3. BINARY OPERATIONS; ADDITION AND MULTIPLICATION IN N.

DEFINITION 3.1 If $A$ is a non-empty set, then a *binary operation,* $\circ$, *on* $A$ is a function $\circ: A \times A \to A$.

*Notation:* If $\circ$ is a binary operation on $A$, we write $a \circ b$ for $\circ(a, b)$, where $a, b \in A$.

DEFINITION 3.2 Let $A$ be a non-empty set, $\circ$ a binary operation on $A$. Then

(1) $\circ$ is commutative if $a \circ b = b \circ a$ for all $a, b \in A$:

(2) $\circ$ is associative if $a \circ (b \circ c) = (a \circ b) \circ c$ for all $a, b, c \in A$.

If $\circ, \circ'$ are binary operations on $A$, then $\circ$ is left-distributive over $\circ'$ if

$$a \circ (b \circ' c) = (a \circ b) \circ' (a \circ c) \text{ for all } a, b, c \in A;$$

and $\circ$ is right-distributive over $\circ'$ if

$$(b \circ' c) \circ a = (b \circ a) \circ' (c \circ a) \text{ for all } a, b, c \in A.$$

We now use the Recursion Theorem to prove that there exist in the natural number system two binary operations with the properties of the addition and multiplication of familiar arithmetic.

THEOREM 3.1 *There is exactly one function* $F: \mathbf{N} \times \mathbf{N} \to \mathbf{N}$ *such that*

(1) $F(m, 1) = S(m)$ *for all* $m \in \mathbf{N}$

(2) $F(m, S(n)) = S(F(m, n))$ *for all* $m, n \in \mathbf{N}$.

PROOF: Let $m$ be any element of $\mathbf{N}$. By the Recursion Theorem, with $A = \mathbf{N}$, $a = S(m)$, and $G = S$, there is exactly one function $F_m: \mathbf{N} \to \mathbf{N}$ such that

(3) $F_m(1) = S(m)$

(4) $F_m(S(n)) = S(F_m(n))$ for all $n \in \mathbf{N}$.

But then the set

$$F = \{((m, n), F_m(n)) \mid (m, n) \in \mathbf{N} \times \mathbf{N}\}$$

is a function from $\mathbf{N} \times \mathbf{N}$ to $\mathbf{N}$. For all $(m, n) \in \mathbf{N} \times \mathbf{N}$, $F(m, n) = F_m(n)$. By (3),

$$F(m, 1) = F_m(1) = S(m) \text{ for all } m \in \mathbf{N}.$$

By (4),

$$F(m, S(n)) = F_m(S(n)) = S(F_m(n)) = S(F(m, n)).$$

Hence $F$ satisfies (1) and (2).

If $\overline{F}$: $\mathbf{N} \times \mathbf{N} \to \mathbf{N}$ satisfies (1) and (2), then

$$\overline{F}(m, 1) = S(m) = F(m, 1) \text{ for each } m \in \mathbf{N}.$$

For all $(m, n) \in \mathbf{N} \times \mathbf{N}$ such that $\overline{F}(m, n) = F(m, n)$, we have

$$\overline{F}(m, S(n)) = S(\overline{F}(m, n)) = S(F(m, n)) = F(m, S(n)).$$

But then for each $m \in \mathbf{N}$, the set $N_m = \{n \in \mathbf{N} \mid \overline{F}(m, n) = F(m, n)\}$ is equal to $\mathbf{N}$ since $N_m$ is an inductive set containing 1. Hence for all $(m, n) \in \mathbf{N} \times \mathbf{N}$, $\overline{F}(m, n) = F(m, n)$, and $F = \overline{F}$.

Since $F$ is a function from $\mathbf{N} \times \mathbf{N}$ to $\mathbf{N}$, it is a binary operation on N.

DEFINITION 3.3 We write "$m + n$" for $F(m, n)$, where $F$ is the function of Theorem 3.3, and use the familiar name "addition" for the binary operation "+."

We can now restate Theorem 3.1.

THEOREM 3.1$^+$ *There is a unique binary operation on* $\mathbf{N}$ *(called addition) such that*

($1^+$) $m + 1 = S(m)$ *for each* $m \in \mathbf{N}$,

($2^+$) $m + S(n) = S(m + n)$ *for each* $m, n \in \mathbf{N}$.

The familiar addition of whole numbers is both associative and commutative. We show that the addition operation we have introduced in the natural number system has both of these properties.

THEOREM 3.2 *Addition in* $\mathbf{N}$ *is associative.*

PROOF: Let $P$ be the set of all $p \in \mathbf{N}$ such that $(m + n) + p = m + (n + p)$ holds for all $m, n \in \mathbf{N}$. Then $1 \in P$, since $(m + n) + 1 = S(m + n) = m + S(n) = m + (n + 1)$, by properties (1) and (2) of addition. If $p \in P$, then $(m + n) + p = m + (n + p)$ for all $m, n \in \mathbf{N}$. Hence $(m + n) + S(p) = S((m + n) + p) = S(m + (n + p)) = m + S(n + p) = m + (n + S(p))$, by ($2^+$). Thus, $P$ is an inductive set containing 1. By the Axiom of Induction, $P = \mathbf{N}$.

THEOREM 3.3 *Addition in* $\mathbf{N}$ *is commutative.*

PROOF: We first show that $m + 1 = 1 + m$ for all $m \in \mathbf{N}$.

Let $M = \{m \mid m + 1 = 1 + m\}$. Then $1 \in M$ since $1 + 1 = 1 + 1$. If $m \in M$, then $S(m) + 1 = (m + 1) + 1 = (1 + m) + 1 = 1 + (m + 1) = 1 + S(m)$ so that $S(m) \in M$. But then $M = \mathbf{N}$. Now let

$$P = \{n \in \mathbf{N} \mid m + n = n + m \text{ for all } m \in \mathbf{N}\}.$$

Then $1 \in P$, since $m + 1 = 1 + m$ for all $m \in \mathbf{N}$. If $n \in P$, then $m + S(n)$

$= m + (n + 1) = m + (1 + n) = (m + 1) + n = n + (m + 1) = n + (1 + m)$
$= (n + 1) + m = S(n) + m$. But then $S(n) \in P$, and, by $A_3, P = \mathbf{N}$.

The associative property of addition was obtained first, and was used in establishing the commutative property. This is not accidental, since associativity was built into the addition operation by imposing conditions $(1^+)$ and $(2^+)$, so that $m + (n + 1) = m + S(n) = S(m + n) = (m + n) + 1$. In introducing a second binary operation, we bear in mind the behavior of 1 in familiar multiplication, and the relationship of multiplication to addition.

THEOREM 3.4 *There is exactly one function* $K\colon \mathbf{N} \times \mathbf{N} \to \mathbf{N}$ *such that*
(1) $K(m, 1) = m$ *for each* $m \in \mathbf{N}$,
(2) $K(m, S(n)) = K(m, n) + m$ *for each* $m, n \in \mathbf{N}$.

PROOF: Let $m$ be any element of $\mathbf{N}$. By the Recursion Theorem, with $A = \mathbf{N}$, $G(k) = k + m$ for each $k \in \mathbf{N}$, and $a = m$, there is exactly one function $K_m\colon \mathbf{N} \to \mathbf{N}$ such that

(3) $K_m(1) = m$,
(4) $K_m(S(n)) = K_m(n) + m$ for all $n \in \mathbf{N}$.

But then the set

$$K = \{((m, n), K_m(n)) \mid (m, n) \in \mathbf{N} \times \mathbf{N}\}$$

is a function from $\mathbf{N} \times \mathbf{N}$ to $\mathbf{N}$. For all $(m, n) \in \mathbf{N} \times \mathbf{N}$, $K(m, n) = K_m(n)$. By (3),

$$K(m, 1) = K_m(1) = m \text{ for all } m \in \mathbf{N}.$$

By (4),

$$K(m, S(n)) = K_m(S(n)) = K_m(n) + m = K(m, n) + m.$$

Hence, $K$ satisfies (1) and (2).

If $\bar{K}$ is any function from $\mathbf{N} \times \mathbf{N}$ to $\mathbf{N}$ satisfying (1) and (2), then

$$\bar{K}(m, 1) = m = K(m, 1) \text{ for all } m \in \mathbf{N}.$$

For all $(m, n) \in \mathbf{N} \times \mathbf{N}$ such that $\bar{K}(m, n) = K(m, n)$ we have

$$\bar{K}(m, S(n)) = \bar{K}(m, n) + m = K(m, n) + m = K(m, S(n)).$$

But then, for each $m \in \mathbf{N}$, the set

$$N_m = \{n \in \mathbf{N} \mid \bar{K}(m, n) = K(m, n)\} = \mathbf{N},$$

since $N_m$ is an inductive set containing 1. Hence for each $(m, n) \in \mathbf{N} \times \mathbf{N}$, $\bar{K}(m, n) = K(m, n)$, and $\bar{K} = K$.

Since $K$ is a function from $\mathbf{N} \times \mathbf{N}$ to $\mathbf{N}$, it is a binary operation on $\mathbf{N}$.

DEFINITION 3.4 We write "$m \cdot n$" or "$mn$" for $K(m, n)$, where $K$ is the function of Theorem 3.4, and use the familiar name "multiplication" for the binary operation "$\cdot$".

We can now restate Theorem 3.4:

THEOREM 3.4$^{\bullet}$ *There is a unique binary operation on* **N** *(called multiplication), such that*

(1$^{\bullet}$) $m \cdot 1 = m$ *for each* $m \in \mathbf{N}$

(2$^{\bullet}$) $m \cdot S(n) = m \cdot n + m$ *for each* $m, n \in \mathbf{N}$.

THEOREM 3.5 *Multiplication in* **N** *is left-distributive over addition, i.e., for all* $m, n, p \in \mathbf{N}$, $m(n+p) = mn + mp$.

PROOF: Let $P$ be the set of all $p \in \mathbf{N}$ such that $m(n+p) = mn + mp$ for all $m, n \in \mathbf{N}$. By (1$^{\bullet}$) and (2$^{\bullet}$), $m(n+1) = m \cdot S(n) = mn + m = mn + m \cdot 1$ for all $m, n \in \mathbf{N}$. Hence, $1 \in P$.

If $p \in P$, then $m(n + S(p)) = m \cdot S(n+p) = m(n+p) + m = (mn + mp) + m = mn + (mp + m) = mn + m \cdot S(p)$, so that $S(p) \in P$.

Hence $P$ is an inductive set containing 1, and $P = \mathbf{N}$.

THEOREM 3.6 *Multiplication in* **N** *is associative.*

PROOF: Let $P$ be the set of all $p \in \mathbf{N}$ such that $(mn)p = m(np)$ for all $m, n \in \mathbf{N}$. By (1$^{\bullet}$), $(mn)1 = mn = m(n1)$ for all $m, n \in \mathbf{N}$. Hence $1 \in P$. If $p \in P$, then $(mn)S(p) = (mn)p + mn = m(np) + mn = m(np + n) = m(nS(p))$, by (2$^{\bullet}$), so that $S(p) \in P$. Hence $P$ is an inductive set containing 1, and $P = \mathbf{N}$.

THEOREM 3.7 *Multiplication in* **N** *is right-distributive over addition, i.e., for all* $m, n, p \in \mathbf{N}$, $(m+n)p = mp + np$.

PROOF: Let $P$ be the set of all $p \in \mathbf{N}$ such that $(m+n)p = mp + np$ for all $m, n \in \mathbf{N}$. Then $(m+n)1 = m + n = m1 + n1$. Hence $1 \in P$. If $p \in P$, then $(m+n)S(p) = (m+n)p + (m+n) = (mp + np) + (m+n) = mp + [np + (m+n)] = mp + [(m+n) + np] = [mp + (m+n)] + np = [(mp + m) + n] + np = (m \cdot S(p) + n) + np = m \cdot S(p) + (n + np) = m \cdot S(p) + (np + n) = m \cdot S(p) + n \cdot S(p)$. Thus, $S(p) \in P$, and $P$ is an inductive set containing 1. But then $P = \mathbf{N}$.

THEOREM 3.8 *Multiplication in* **N** *is commutative.*

PROOF: We first show that $m \cdot 1 = 1 \cdot m$ for all $m \in \mathbf{N}$. Let

$$M = \{m \in \mathbf{N} \mid m \cdot 1 = 1 \cdot m\}.$$

Then $1 \in M$, since $1 \cdot 1 = 1 \cdot 1$. If $m \in M$, then

$$1 \cdot S(m) = 1(m + 1) = 1 \cdot m + 1 = m \cdot 1 + 1 = m + 1 = S(m)$$

and, by the definition of multiplication, $S(m) \cdot 1 = S(m)$. Hence $S(m) \in M$, and so $M = \mathbf{N}$. Now let

$$P = \{n \in \mathbf{N} \mid m \cdot n = n \cdot m \text{ for all } m \in \mathbf{N}\}.$$

Then $1 \in P$, since $m \cdot 1 = 1 \cdot m$ for all $m \in \mathbf{N}$. If $n \in P$, then

$$m \cdot S(n) = mn + m = nm + 1m = (n + 1)m = S(n)m.$$

Hence, $S(n) \in P$ and $P = \mathbf{N}$.

*Exercise 3.1* For $m, n \in \mathbf{N}, m + n \neq n$.

*Exercise 3.2* For all $m, n \in \mathbf{N}, m + n \neq m$.

*Exercise 3.3* Define $2 = S(1)$, $3 = S(2)$, $4 = S(3)$, $5 = S(4)$, $6 = S(5)$, and prove

$$1 + 1 = 2,$$
$$2 + 4 = 3 + 3 = 6.$$

*Exercise 3.4* Let $A$ be a non-empty set, $M_A$ the set of all functions from $A$ to $A$, $S_A$ the set of all bijections from $A$ to $A$.

(a) Note that "composition of functions" is an associative binary operation on $M_A$. (See Th. 0, 4.1.)

(b) Prove that composition is non-commutative except when $A$ is a singleton.

(c) Prove that the restriction of composition to $S_A \times S_A$ is a binary operation on $S_A$.

**4. ORDER IN N.** If one whole number is less than another, then the second can be obtained by adding a whole number to the first. For the set $\mathbf{N}$ of natural numbers we have

THEOREM 4.1 *If $T$ is the subset of $\mathbf{N} \times \mathbf{N}$ consisting of all $(m, n)$ such that $m + p = n$ for some $p \in \mathbf{N}$, then $T$ is an order relation in $\mathbf{N}$.*

PROOF: Since $T$ is a subset of $\mathbf{N} \times \mathbf{N}$, it is a binary relation in $\mathbf{N}$. To show that $T$ is an order relation (Definition 0, 3.6) we show that

(1) If $m, n \in \mathbf{N}$, then one and only one of the statements

$$m = n \qquad (m, n) \in T \qquad (n, m) \in T$$

is true (trichotomy).

(2) If $(m, n) \in T$ and $(n, p) \in T$, then $(m, p) \in T$ (transitivity).

For each $m \in \mathbf{N}$, let $M_m$ be the set of all $n \in \mathbf{N}$ such that one of the statements in (1) is true. We show that $M_m = \mathbf{N}$ for all $m \in \mathbf{N}$. By Theorem 1.1, either $m = 1$ or $m = p + 1$ for some $p \in \mathbf{N}$. Hence either $m = 1$, or $(1, m) \in T$. In either case $1 \in M_m$. Now assume $n \in M_m$. If $m = n$, then $S(n) = m + 1$. Hence, $(m, S(n)) \in T$ and $S(n) \in M_m$. If $(m, n), \in T$, then $m + p = n$ for some $p \in \mathbf{N}$ and $m + S(p) = S(n)$. Hence, $(m, S(n)) \in T$ and $S(n) \in M_m$. If $(n, m) \in T$, then $n + p = m$ for some $p \in \mathbf{N}$. But $p = 1$ or $p = S(q)$ for some $q \in \mathbf{N}$. In the first case, $S(n) = m$, while in the second case, $S(n) + q = n + S(q) = m$ and $(S(n), m) \in T$. In either case, $S(n) \in M_m$. By the Axiom of Induction, $M_m = \mathbf{N}$ for all $m \in \mathbf{N}$. Now if $m, n \in \mathbf{N}$, then $n \in M_m$ and at least one of the statements of (1) is true.

It remains to be shown that for $m, n \in \mathbf{N}$ not more than one of the statements of (1) is true. If $m = n$ and $(m, n) \in T$, then $m + p = n = m$ for some $p \in \mathbf{N}$. If $m = n$ and $(n, m) \in T$, then $n + q = m = n$ for some $q \in \mathbf{N}$. If $(m, n) \in T$ and $(n, m) \in T$, then $m + p = n$ and $n + q = m$ for some $p, q \in \mathbf{N}$. Hence, $m + (p + q) = (m + p) + q = n + q = m$. In each case, we have a contradiction, since $m + t \neq m$ for $m, t \in \mathbf{N}$ (Exercise 3.2). Hence not more than one of the statements in (1) is true for any $m, n \in \mathbf{N}$. This completes the proof of (1).

By the hypothesis of (2) there are $r, s \in \mathbf{N}$ such that

$$m + r = n, \qquad n + s = p.$$

Hence,

$$m + (r + s) = (m + r) + s = n + s = p$$

and $(m, p) \in T$. This proves (2).

DEFINITION 4.1 We write "$m < n$" ("$n > m$") for "$(m, n) \in T$," where $T$ is the order relation of Theorem 4.1, and we read "$m$ is less than $n$" ("$n$ is greater than $m$"). If $n = m + p$, we write "$p = n - m$" and observe that "$n - m$" is defined for $m, n \in \mathbf{N}$ only if $m < n$.

We can now restate Theorem 4.1:

(1) If $m, n \in \mathbf{N}$, then just one of $m = n, m < n, n < m$ ($m = n, n > m, m > n$) is true (trichotomy).

(2) If $m < n$ and $n < p$ then $m < p$ (if $n > m$ and $p > n$, then $p > m$) (transitivity).

The following theorem states that the order in $\mathbf{N}$ is not disturbed by addition or multiplication.

THEOREM 4.2 *For $m, n, p \in \mathbf{N}$, $m < n$ implies $m + p < n + p$ and $mp < np$.*

We leave the proof as an exercise. (Exercise 4.1).

The order in **N** may be used to define subsets of **N**. Examples: The set of all $n \in \mathbf{N}$ such that $S(1) < n$; the set of all $n \in \mathbf{N}$ such that $n < S(m)$ for some $m \in \mathbf{N}$. It is convenient to introduce notation for stating such definitions.

*Notation:* We write

"$m \leqq n$" for "$m < n$ or $m = n$,"
"$m < n < p$" for "$m < n$ and $n < p$,"
"$m \leqq n < p$" for "$m \leqq n$ and $n < p$,"
"$m < n \leqq p$" for "$m < n$ and $n \leqq p$,"
"$m \leqq n \leqq p$" for "$m \leqq n$ and $n \leqq p$."

*Remark:* "$\leqq$" is a partial order relation in **N** in the sense of Definition 0, 3.5. (See Exercise 4.3).

A first element for a subset of **N** is defined as follows:

DEFINITION 4.2 If $M \subset \mathbf{N}$ and there is some $p \in M$ such that

$$p \leqq m \text{ for all } m \in M,$$

then $p$ is called a *first element* (*least element*) of $M$.

*Remark:* If $M \subset \mathbf{N}$ and $p$ and $q$ are first elements of $M$, then $p = q$. (See Exercise 4.4.)

Thus, if $M$ has a first element, it has only one first element. We shall speak of *the* first element of $M$.

Some familiar examples of ordered sets do not have first elements. The present moment is earlier than all future moments. The set of all future moments does not have an earliest moment. We shall see that every non-empty subset of **N** has a first element.

THEOREM 4.2 *1 is the first element of* **N**.

PROOF: If $m \in \mathbf{N}$, then $m = 1$ or $m = S(p) = p + 1$ for some $p \in \mathbf{N}$. If $m = p + 1$, then $1 < m$ by Definition 4.1. Hence, $1 \leqq m$ for all $m \in \mathbf{N}$. By Definition 4.2 and Exercise 4.4, 1 is the (unique) first element of **N**.

COROLLARY *If $M$ is a subset of* **N** *such that $1 \in M$, then 1 is the first element of $M$.*

THEOREM 4.3 *If $n \in \mathbf{N}$, then the set of all $m \in \mathbf{N}$ such that $n < m < S(n)$ is empty.*

PROOF: If $n < m$ in $\mathbf{N}$, then there is some $p \in \mathbf{N}$ such that

$$m = n + p.$$

By Theorem 1, 1.1, $p = 1$ or $p = S(q)$ for some $q \in \mathbf{N}$. If $p = 1$ then

$$m = n + 1 = S(n),$$

and $m < S(n)$ is false by trichotomy. If $p = S(q)$ then

$$m = n + (q + 1) = (n + 1) + q \text{ and } S(n) < m.$$

Hence, again by trichotomy, $m < S(n)$ is false. But then the set of all $m \in \mathbf{N}$ such that $n < m < S(n)$ is empty.

DEFINITION 4.3 For $n \in \mathbf{N}$, the *initial segment* $I_n$ is the set of all $m \in \mathbf{N}$ such that $m \leqq n$.

THEOREM 4.4 *If $M$ is a subset of $\mathbf{N}$ such that*
(1) $I_1 \subset M$,
(2) $S(n) \in M$ *whenever* $I_n \subset M$,
*then* $M = \mathbf{N}$.

PROOF: By the definition of $I_n$ and Theorem 4.3,

$$I_{S(n)} = I_n \cup \{S(n)\}.$$

Hence, by (2),

(3) If $I_n \subset M$ then $I_{S(n)} \subset M$.

By (1), (3), and the Axiom of Induction, $I_n \subset M$ for all $n \in \mathbf{N}$. Since $n \in I_n$, $\mathbf{N} \subset M$ and, since $M \subset \mathbf{N}$ by hypothesis, $M = \mathbf{N}$.

Since the hypothesis of (2) in Theorem 4.4 asserts "more" than the hypothesis "$n \in M$" of (2) in the Principle of Induction, the hypothesis—consisting of (1) and (2)—of Theorem 4.4 asserts "less" than the hypothesis of the Principle of Induction. But the conclusion, "$M = \mathbf{N}$," of Theorem 4.4 is also the conclusion of the Principle of Induction. In this sense Theorem 4.4 is "stronger" than the Principle of Induction. Theorem 4.4 is sometimes referred to as the Second Principle of Induction.

THEOREM 4.5 *Every non-empty subset of $\mathbf{N}$ contains a first element.*

PROOF: We prove the equivalent statement: If $M \subset \mathbf{N}$ and $M$ contains no first element, then $M$ is empty. Let $K$ be the set of all natural numbers not in $M$. Since $1 \leqq n$ for all $n \in \mathbf{N}$ and $M$ contains no first element,

(1) $1 \in K$.

Further,

(2) If $I_n \subset K$, then $S(n) \in K$.

For, if $p \in M$, then $p \notin I_n$, since $I_n \subset K$. Hence, $n < p$ and, by Theorem 4.3, $S(n) \leqq p$. Since $M$ contains no first element, $S(n) \notin M$. Hence, $S(n) \in K$.

By (1), (2), and the Second Principle of Induction (Theorem 4.4), $K = \mathbf{N}$, and $M$ is empty.*]*

*Exercise 4.1* Prove Theorem 4.2: for $m, n, p \in \mathbf{N}$, $m < n$ implies $m + p < n + p$ and $mp < np$.

*Exercise 4.2* Let $m, n, p \in \mathbf{N}$. Use Exercise 4.1 and the trichotomy of $<$ to prove

(a) If $m + p < n + p$, then $m < n$;
(b) if $mp < np$, then $m < n$;
(c) if $m + p = n + p$, then $m = n$;
(d) if $mp = np$, then $m = n$.

*Exercise 4.3* Prove that the relation $\leqq$ is a partial order in $\mathbf{N}$.

*Exercise 4.4* Prove that, if $M \subset \mathbf{N}$ and $p, q \in M$ are both first elements of $M$, then $p = q$.

*Exercise 4.5* Prove: for each $m, p \in \mathbf{N}$, $m \leqq mp$.

**5. GENERALIZED ASSOCIATIVE AND COMMUTATIVE LAWS.** To simplify our work with sums and products in $\mathbf{N}$ and in the systems we introduce later, we define a composite of $n$ elements in an arbitrary semigroup $G$ and prove the generalized associative law, and, in case the operation in $G$ is commutative, the generalized commutative law. These theorems will allow us to rearrange and reassociate freely the terms in sums and the factors in products.

DEFINITION 5.1 Let $G$ be a set, $\circ$ an associative binary operation on $G$ and $(b_h)$ a sequence of elements in $G$. Then the *n-composite* $\underset{h=1}{\overset{n}{\mathsf{P}}} b_h$ is defined* by

$$\underset{h=1}{\overset{1}{\mathsf{P}}} b_h = b_1$$

$$\underset{h=1}{\overset{n+1}{\mathsf{P}}} b_h = \left(\underset{h=1}{\overset{n}{\mathsf{P}}} b_h\right) \circ b_{n+1}.$$

*This definition is an application of the Generalized Recursion Theorem. What are $A$, $a$, the mappings $G_n$ and $F$ in this case? (cf. page 22).

If, for $b \in G$, $b_h = b$ for all $h \leqq n$, we write $b^n$ for $\underset{h=1}{\overset{n}{\mathrm{P}}} b_h$. When the usual notations "+" and "·" are used for addition and multiplication, P will be replaced by $\Sigma$ or $\Pi$, respectively.

THEOREM 5.1 (GENERALIZED ASSOCIATIVE LAW) *Let G be a set,* · *an associative binary operation on G and* $(b_h)$ *a sequence of elements in G. For* $n \in \mathbf{N}$, $k \in I_n$, *let* $F = F_k^n$ *be a mapping of* $I_k$ *into* $I_n$ *such that*

$$(1)\quad n_1 < n_2 < \cdots < n_k = n,$$

*where*

$$n_j = F(j) \text{ for } j = 1, 2, \ldots, k.$$

Then

$$(2)\quad \underset{h=1}{\overset{n}{\mathrm{P}}} b_h = \underset{j=1}{\overset{k}{\mathrm{P}}} Q_j,$$

*where*

$$Q_1 = \underset{h=1}{\overset{n_1}{\mathrm{P}}} b_h$$

*and*

$$Q_j = \underset{h=n_{j-1}+1}{\overset{n_j}{\mathrm{P}}} b_h \text{ for } j = 2, \ldots, k.$$

PROOF: Let $M$ be the set of all $n \in \mathbf{N}$ such that (2) holds for all $k \in I_n$ and all mappings $F_k^n$ satisfying (1).

Now, $1 \in M$, since from $n = 1$ it follows that $k = 1$, and $F = F_1^1$ is the identity mapping on $I_1 = \{1\}$. Hence

$$\underset{h=1}{\overset{n}{\mathrm{P}}} b_h = \underset{h=1}{\overset{1}{\mathrm{P}}} b_h = b_1 = Q_1 = \underset{j=1}{\overset{1}{\mathrm{P}}} Q_j = \underset{j=1}{\overset{k}{\mathrm{P}}} Q_j.$$

Suppose $I_n \subset M$, and $F_k^{n+1}$ is any mapping of $I_k$ into $I_{n+1}$ satisfying (1). If $k = 1$, then $n_1 = n + 1$, and

$$\underset{j=1}{\overset{k}{\mathrm{P}}} Q_j = Q_1 = \underset{h=1}{\overset{n_1}{\mathrm{P}}} b_h$$

so that (2) is satisfied and $n + 1 \in M$. If $k > 1$, then

$$Q_k = \underset{h=n_{k-1}+1}{\overset{n+1}{\mathrm{P}}} b_h = \left( \underset{h=n_{k-1}+1}{\overset{n}{\mathrm{P}}} b_h \right) \circ b_{n+1}.$$

Since $n_{k-1} \in I_n$,

$$\operatorname*{P}_{j=1}^{k} Q_j = \left(\operatorname*{P}_{j=1}^{k-1} Q_j\right) \circ Q_k = \left(\operatorname*{P}_{h=1}^{n_{k-1}} b_h\right) \circ \left[\left(\operatorname*{P}_{h=n_{k-1}+1}^{n} b_h\right) \circ b_{n+1}\right].$$

But then

$$\operatorname*{P}_{j=1}^{k} Q_j = \left(\operatorname*{P}_{h=1}^{n_{k-1}} b_h \circ \operatorname*{P}_{h=n_{k-1}+1}^{n} b_h\right) \circ b_{n+1},$$

by the associativity of the operation in $G$. Since $n \in M$,

$$\operatorname*{P}_{j=1}^{k} Q_j = \left(\operatorname*{P}_{h=1}^{n} b_h\right) \circ b_{n+1} = \operatorname*{P}_{h=1}^{n+1} b_h, \text{ by Definition 5.1}$$

Thus, (2) is satisfied and $n+1 \in M$. By the second induction principle, $M = \mathbf{N}$.

THEOREM 5.2 (GENERALIZED COMMUTATIVE LAW) *Let $G$ be a set, $\circ$ an associative and commutative binary operation on $G$ and $(b_h)$ a sequence of elements in $G$. For $n \in \mathbf{N}$, let $F = F_n$ be a* 1-1 *mapping of $I_n$ onto itself. Then*

$$(1) \quad \operatorname*{P}_{h=1}^{n} b_n = \operatorname*{P}_{h=1}^{n} b_{n_h} \textit{ where } n_h = F(h) \text{ for } h \in I_n.$$

PROOF: Let $M$ be the set of all $n \in \mathbf{N}$ such that (1) holds for each 1-1 mapping $F$ of $I_n$ onto itself.

If $n = 1$, then $I_n = I_1$ and $F$ is the identity mapping on $I_1$. Hence, $\operatorname*{P}_{h=1}^{1} b_h = b_1 = b_{n_1} = \operatorname*{P}_{h=1}^{1} b_{n_h}$, and $1 \in M$. Suppose $n \in M$, and $F = F_{n+1}$ is a 1-1 mapping of $I_{n+1}$ onto itself. If $b_{n_{n+1}} = b_{n+1}$, then, since $n \in M$,

$$\operatorname*{P}_{h=1}^{n+1} b_h = \left(\operatorname*{P}_{h=1}^{n} b_h\right) \circ b_{n+1} = \left(\operatorname*{P}_{h=1}^{n} b_{n_h}\right) \circ b_{n_{n+1}} = \operatorname*{P}_{h=1}^{n+1} b_{n_h},$$

so that $n+1 \in M$. If $b_{n+1} = b_{n_l}$ for $1 \leqq l \leqq n$, then $\operatorname*{P}_{h=1}^{n+1} b_{n_h} = \operatorname*{P}_{h=1}^{l} b_{n_h} \operatorname*{P}_{h=l+1}^{n+1} b_{n_h}$, by Theorem 5.1. Since multiplication in $G$ is commutative,

$$\operatorname*{P}_{h=1}^{n+1} b_{n_h} = \operatorname*{P}_{h=l+1}^{n+1} b_{n_h} \circ \operatorname*{P}_{h=1}^{l} b_{n_h}.$$

If $l = 1$, then

$$\operatorname*{P}_{i=1}^{n+1} b_{n_h} = \left(\operatorname*{P}_{h=2}^{n+1} b_{n_h}\right) \circ b_{n_1} = \left(\operatorname*{P}_{h=1}^{n} b_{n'_h}\right) \circ b_{n+1}.$$

where $n'_h = n_{h+1}$ for $h = 1, \ldots, n$. Since $n \in M$,

$$\mathop{\mathrm{P}}_{h=1}^{n+1} b_{n_h} = \left( \mathop{\mathrm{P}}_{h=1}^{n} b_h \right) \circ b_{n+1} = \mathop{\mathrm{P}}_{h=1}^{n+1} b_h .$$

If $l > 1$, then

$$\mathop{\mathrm{P}}_{h=1}^{n+1} b_{n_h} = \mathop{\mathrm{P}}_{h=l+1}^{n+1} b_{n_h} \circ \left( \mathop{\mathrm{P}}_{h=1}^{i-1} b_{n_h} b_{n+1} \right)$$

$$= \left( \mathop{\mathrm{P}}_{h=l+1}^{n+1} b_{n_h} \circ \mathop{\mathrm{P}}_{h=1}^{l-1} b_{n_h} \right) \circ b_{n+1} = \left( \mathop{\mathrm{P}}_{h=1}^{l-1} b_{n_h} \circ \mathop{\mathrm{P}}_{h=1}^{n} b_{n'_h} \right) \circ b_{n+1} .$$

*Exercise 5.1* For each $n \in \mathbf{N}$, let $\sigma_n = \sum_{i=1}^{n} i$. Prove:

(a) $\sigma_{n+1} = \sigma_n + n + 1$ for each $n \in \mathbf{N}$.
(b) $\sigma_n \geqslant n$ for each $n \in \mathbf{N}$.
(c) If $m < n$, then $\sigma_m < \sigma_n$.
(d) If $n \in \mathbf{N}, n > 1$, then there is a unique natural number $m$ such that

$$\sigma_m < n \leqslant \sigma_{m+1} .$$

**6. COUNTING.** The familiar process of counting uses the whole numbers as a standard set of tags. Formally, we shall define a non-empty set to be finite if it can be tagged with the numbers in the initial segment $I_n$, for some $n \in \mathbf{N}$, and countable if it can be tagged with all of the natural numbers. We shall see that some sets are neither finite nor countable and that, in fact, given any set, there is a "larger set."

First, we must find a means of comparing sets for size. Our experience suggests that two sets $A$ and $B$ are "the same size" if the elements of $A$ can be put in one-to-one correspondence with the elements of $B$. If this cannot be done, then one of the sets is "smaller" than the other, and the "smaller set" is the same size as a subset of the "larger."

DEFINITION 6.1 Let $A$, $B$ be non-empty sets. Then $A$ *is equipotent to* $B$ ($A \sim B$) if there is a bijection from $A$ to $B$; $A$ *is dominated by* $B$ ($A \lesssim B$) if there is an injection from $A$ to $B$; and $A$ *is strictly dominated by* $B$ ($A \prec B$) if there is an injection from $A$ to $B$, but there is no bijection from $A$ to $B$.

*Remarks:* 1. On any non-empty set of sets, $\sim$ is an equivalence relation. (See Exercise 6.1.)

2. $A \lesssim B$ if and only if $A \prec B$ or $A \sim B$. (See Exercise 6.2)

THEOREM 6.1 *If $X$ is a non-empty set, then*
(1) *there is no surjection from $X$ to $\mathscr{P}(X)$;*
(2) $X \prec \mathscr{P}(X)$.

PROOF: (1) Suppose there is a surjection $F: X \to \mathscr{P}(X)$. Let $Y = \{x \in X | x \notin F(x)\}$. Since $Y \in \mathscr{P}(X)$ and $F$ is surjective, there is an element $y \in X$ such that $Y = F(y)$. Then $y \in Y$ if and only if $y \notin F(y)$. This is impossible since $Y = F(y)$. Hence there is no surjection from $X$ to $\mathscr{P}(X)$.

(2) If $G: X \to \mathscr{P}(X)$ is defined by $G(x) = \{x\}$ for each $x \in X$, then $G$ is an injection, by the Axiom of Identity. By (1), there is no surjection, hence certainly no bijection, from $X$ to $\mathscr{P}(X)$. It follows that $X \prec \mathscr{P}(X)$.

The preceding theorem indicates that, for any set, there is a "larger" set. A theory of cardinal numbers may be developed to reflect the hierarchy of "sizes" of sets. In this book, we confine ourselves to a discussion of the concepts introduced in the following definition.

DEFINITION 6.2 Let $A$ be a set. Then
(1) $A$ is *finite* if $A = \phi$ or $A \sim I_n$ for some $n \in \mathbf{N}$;
(2) $A$ is *countable* if $A \sim \mathbf{N}$;
(3) $A$ is *infinite* if there is a proper subset $B$ of $A$ such that $A \sim B$.

COROLLARY *A set $A$ is infinite if and only if there is an injection $F: A \to A$ such that $F(A) \neq A$.*

PROOF: If $A$ is infinite, then there is a bijection $\overline{F}: A \to B \subsetneqq A$. The function $F: A \to A$ such that $F(a) = \overline{F}(a)$ for each $a \in A$ is an injection such that $F(A) = B \neq A$.

Conversely, if $F: A \to A$ is an injection such that $F(A) \neq A$, then $B = F(A)$ is a proper subset of $A$ which is equipotent to $A$ under the bijection $\overline{F}: A \to B$ such that $\overline{F}(a) = F(a)$ for each $a \in A$. Hence $A$ is infinite.

*Example 1:* For each $n \in \mathbf{N}$, the initial segment, $I_n$, is finite since $I_n \sim I_n$, under the identity function on $I_n$.

*Example 2:* $\mathbf{N}$ is countable since $\mathbf{N} \sim \mathbf{N}$, under the identity function on $\mathbf{N}$.

*Example 3:* $\mathscr{P}(\mathbf{N})$ is not countable since, by Theorem 6.1, $\mathbf{N} \prec \mathscr{P}(\mathbf{N})$, hence $\mathbf{N}$ is not equipotent to $\mathscr{P}(\mathbf{N})$ (see Definition 6.1). But then $\mathscr{P}(\mathbf{N})$ is not equipotent to $\mathbf{N}$ (see Exercise 6.1).

*Example 4:* $\mathbf{N}$ is infinite since $S: \mathbf{N} \to \mathbf{N}$ is an injection such that $S(\mathbf{N}) \neq \mathbf{N}$ (see Corollary, Definition 6.2). Note that $\mathbf{N} \sim S(\mathbf{N})$, a proper subset of $\mathbf{N}$.

Each of the properties introduced in Definition 6.2 is "invariant under equipotence," i.e., if $A \sim B$ and $A$ is finite, then $B$ is finite; if $A \sim B$ and $A$ is countable, then $B$ is countable; if $A \sim B$ and $A$ is infinite, then $B$ is infinite. We leave the proof of the first two assertions as an exercise (Exercise 6.3) and prove the third.

THEOREM 6.2 *If $A$, $B$ are sets such that $A \sim B$ and $A$ is infinite, then $B$ is infinite.*

PROOF: Since $A \sim B$, there is a bijection $F: A \to B$. Since $A$ is infinite, there is an injection $G: A \to A$ such that $G(A) \neq A$. (See Corollary, Definition 6.2). Let $H = FGF^{-1}$. Then $H$ is an injection from $B$ to $B$. We show that $H(B) \neq B$. Since $G(A) \neq A$, there is some $a \in A - G(A)$. Suppose $F(a) \in H(B)$. Then, for some $b \in B$, $F(a) = H(b) = FGF^{-1}(b)$. But then, since $F$ is injective, $a = GF^{-1}(b) = G(F^{-1}(b)) \in G(A)$, in contradiction to the choice of $a$. It follows that $H(B) \neq B$. But then $B$ is equipotent to its proper subset, $H(B)$, and so $B$ is infinite.

From Definition 6.2, it follows immediately that $\mathbf{N}$ is countable, and that $I_n$ is finite for each $n \in \mathbf{N}$. We now investigate whether "infinite" and "not finite," in the sense of Definition 6.2, are equivalent.

THEOREM 6.3 *For each $n \in \mathbf{N}$, the initial segment $I_n$ is not infinite.*

PROOF: Let $M = \{n \in \mathbf{N} | I_n \text{ is not infinite}\}$. $1 \in M$, for: $I_1 = \{1\}$ has no proper subsets. Suppose $n \in M$. If $S(n) \notin M$, then there is a bijection $F: I_{S(n)} \to K$, where $K$ is a proper subset of $I_{S(n)}$.

*Case 1* $S(n) \in K$ and $F(S(n)) = S(n)$. Let $K' = K - \{S(n)\}$. Since $K \sim I_{S(n)}$, $K$ is not a singleton, hence $K' \neq \phi$. Also, since $K$ is a proper subset of $I_{S(n)}$, $K'$ is a proper subset of $I_n$. The set $F' = F - \{(S(n), F(S(n)))\}$ defines a bijection from $I_n$ to $K'$. But then $I_n$ is infinite, contrary to the hypothesis that $n \in M$.

*Case 2* $S(n) \in K$ and $F(S(n)) \neq S(n)$. Then there is some $k \in K$, $k \neq S(n)$, such that $F(S(n)) = k$, and there is some $t \in I_{S(n)}$, $t \neq S(n)$, such that $F(t) = S(n)$. Swap images for $t$ and $S(n)$, i.e., define $F'': I_{S(n)} \to K$ by: $F''(i) = F(i)$ for $i \neq t$, $i \neq S(n)$; $F''(t) = k$ and $F''(S(n)) = S(n)$. Then $F'': I_{S(n)} \to K$ is a bijection satisfying the conditions imposed on $F$ in Case 1, and we have shown that Case 1 leads to a contradiction.

*Case 3* $S(n) \notin K$. Then $K \subset I_n$ and $F'' = F - \{(S(n), F(S(n)))\}$ defines a bejection of $I_n$ onto $K' = K - \{F(S(n))\}$, a proper subset of $I_n$, contrary to the hypothesis that $n \in M$.

Hence $S(n) \in M$ and the theorem follows by the Principle of Induction.

COROLLARY *If $X$ is a finite set, then $X$ is not infinite.*

PROOF: Let $X$ be a finite set. Then $X = \phi$ or $X \sim I_n$ for some $n \in \mathbf{N}$. If $X = \phi$, then $X$ is not infinite since $\phi$ has no proper subsets. If $X \sim I_n$ and $X$ is infinite, then $I_n$ is infinite, by Theorem 6.1. Contradiction! Hence $X$ is not infinite.

THEOREM 6.4 *If $X$ is a set which is not finite, then $X$ has a countable subset.*

PROOF: Let $a = \{Y \in P(X) \mid Y \text{ is finite}\}$. Since $X$ is not finite, $X \neq \phi$ and, since the finite set $\phi$ is an element of $P(X)$, $\phi \in a$. If $Y \in a$, then $Y \subset X$ and $X - Y \neq \phi$.

By the Axiom of Choice, there is a choice function $T\colon P(X) - \{\phi\} \to X$ such that $T(Y) \in Y$ for each $Y \in P(X) - \{\phi\}$. In particular, $T(X - Y) \in X - Y$ for each $Y \in a$. Also, for $Y \in a$, $Y \cup \{T(X - Y)\} \in a$ (see Exercise 6.5).

Define $G\colon a \to a$ by: $G(Y) = Y \cup \{T(X - Y)\}$, for each $Y \in a$. By the Recursion Theorem (Th. 2.1), there is a function $F\colon \mathbf{N} \to a$ such that $F(1) = \phi$ and $F(S(n)) = G(F(n)) = F(n) \cup \{T(X - F(n))\}$ for each $n \in \mathbf{N}$. Now, $F(n) \subset F(S(n))$ for each $n \in \mathbf{N}$ and, by induction, $F(n) \subset F(m)$ if $n \leqslant m$ in $\mathbf{N}$. Also, from $n < m$ in $\mathbf{N}$, it follows that $S(n) \leqslant m$, hence $F(S(n)) \subset F(m)$, and

$$(1) \qquad T(X - F(n)) \in F(S(n)) \subset F(m),$$

$$T(X - F(m)) \notin F(m).$$

Thus, if $n \neq m$ in $\mathbf{N}$, then $n < m$ or $m < n$ and, by (1), $T(X - F(n)) \neq T(X - F(m))$. Let $C = \{T(X - F(n)) \mid n \in \mathbf{N}\} \subset X$. Then $H\colon \mathbf{N} \to C$ such that $H(n) = T(X - F(n))$ for each $n \in \mathbf{N}$ is a bijection, hence $C$ is a countable subset of $X$.

COROLLARY *If $X$ is a set which is not finite, then $N \lesssim X$.*

THEOREM 6.5 *If a set is not finite, then it is infinite.*

PROOF: Let $X$ be a set which is not finite. By Theorem 6.4, there is a bijection $H\colon \mathbf{N} \to C$, where $C$ is a subset of $X$. Define $F\colon X \to X$ by: $F(x) = HSH^{-1}(x)$ for $x \in C$,

$$F(x) = x \qquad \text{for } x \in X - C.$$

Then $F$ is an injection, for: if $x, x' \in C$ and $F(x) = F(x')$, then $HSH^{-1}x = HSH^{-1}x'$, hence $x = x'$, since $HSH^{-1}$ is an injection. If $x, x' \in X - C$ and $F(x) = F(x')$, then $x = F(x) = F(x') = x'$. And, for $x \in C$, $x' \in X - C$, $F(x) = F(x')$ is impossible since $F(x) = HSH^{-1}x \in H(\mathbf{N}) = C$, while $F(x')$ $x' \in X - C$. Thus, $F$ is an injection.

But $F$ is not a surjection, for: if $H(1) \in F(X)$, then $H(1) = F(x)$ for some

$x \in C$. Hence $H(1) = HSH^{-1}(x)$, and $1 = SH^{-1}(x) = S(H^{-1}(x))$. This is impossible since 1 is not a successor.

Thus, $F\colon X \to X$ is an injection such that $F(X) \neq X$ By the Corollary of Definition 6.2, it follows that $X$ is infinite.

COROLLARY *A set X is infinite if and only if it is not finite.*

PROOF: If $X$ is infinite, then $X$ is not finite, by the Corollary of Theorem 6.2. If $X$ is not finite, then $X$ is infinite, by Theorem 6.5.

THEOREM 6.6 $\mathbf{N} \times \mathbf{N}$ *is countable.*

PROOF: We list the elements of $\mathbf{N} \times \mathbf{N}$ as indicated in the diagram below, then count them by following the arrows.

$$
\begin{array}{c}
(1, 1) \\
\swarrow \\
(2, 1) \to (1, 2) \\
\swarrow \\
(3, 1) \to (2, 2) \to (1, 3) \\
\swarrow \\
(4, 1) \to (3, 2) \to (2, 3) \to (1, 4) \\
\cdot \qquad\qquad\qquad \cdot \\
\cdot \qquad\qquad\qquad\quad \cdot \\
\cdot \qquad\qquad\qquad\qquad \cdot
\end{array}
$$

Define a sequence $(\sigma_m)$ as follows: $\sigma_m = \sum_{i=1}^{m} i$ for each $m \in \mathbf{N}$. Then, for each $n \in \mathbf{N}, n > 1$, there are unique natural numbers $m$ and $q$ such that $n = \sigma_m + q$, $1 \leqslant q \leqslant m + 1$ (see Exercise 5.1).

Now define $F\colon \mathbf{N} \to \mathbf{N} \times \mathbf{N}$ by:

$$F(1) = (1, 1)$$

$$F(n) = (m + 2 - q, q) \text{ if } n = \sigma_m + q,\ 1 \leqslant q \leqslant m + 1.$$

We prove that $F$ is a bijection by producing $F^{-1}$. (See Theorem 0, 4.2) Define $G\colon \mathbf{N} \times \mathbf{N} \to \mathbf{N}$ as follows: $G(1, 1) = 1$ and, for $(p, q) \neq (1,1)$, $G(p, q) = \sigma_{p+q-2} + q$. Let $n \in \mathbf{N}$. For $n = 1$, $G(F(1)) = G(1, 1) = 1$. For $n > 1$, we have $n = \sigma_m + q$ for some $m, q \in \mathbf{N}$, $1 \leqslant q \leqslant m + 1$, hence $G(F(n)) = G(m + 2 - q, q) = \sigma_{m+2-q+q-2} + q = \sigma_m + q = n$. Thus, $GF = I_{\mathbf{N}}$.

Next, let $(p, q) \in \mathbf{N} \times \mathbf{N}$. If $(p, q) = (1, 1)$, then $F(G(1, 1)) = F(1) = (1, 1)$. If $(p, q) \neq (1, 1)$, then $F(G(p, q)) = F(\sigma_{p+q-2} + q) = (p + q - 2 + 2 - q, q) = (p, q)$. Thus, $FG = I_{\mathbf{N} \times \mathbf{N}}$.

But then $F$ is a bijection, hence $\mathbf{N} \times \mathbf{N} \sim \mathbf{N}$, i.e., $\mathbf{N} \times \mathbf{N}$ is countable.

COROLLARY *Let $A$ be a countable set. Then $A \times A$ is countable. We leave the proof as an exercise. (Exercise 6.19.)*

THEOREM 6.7 *Every subset of* **N** *is either finite or countable.*

PROOF: Let $A$ be a subset of **N**. Suppose $A$ is not finite. Since every subset of a finite set is finite (see Exercise 6.10), $A$ is not a subset of any initial segment of **N**, i.e., for each $n \in \mathbf{N}$, $A - I_n \neq \phi$. A particular choice function for $A$ is the "least element function" $l: \mathscr{P}(A) - \{\phi\} \to A$ which associates with each non-empty subset $B$ of $A$ its least element, $l(B)$.

Define $G: A \to A$ by: $G(m) = l(A - I_m)$ for each $m \in \mathbf{N}$. By the Recursion Theorem, there is a function $F: \mathbf{N} \to A$ such that $F(1) = l(A)$ and $F(S(n)) = G(F(n)) = l(A - I_{F(n)})$ for each $n \in \mathbf{N}$. We prove that $F$ is a bijection.

Since $F(S(n)) = l(A - I_{F(n)}) \in A - I_{F(n)}$, we have $F(n) < F(S(n))$ for each $n \in \mathbf{N}$. By induction, it follows that, for $m, n \in \mathbf{N}$, $m < n$ implies $F(m) < F(n)$. Thus, $F$ is injective.

Let $k \in A$. Since $F$ is injective, $F(\mathbf{N})$ is equipotent to **N**, hence infinite. But then $F(\mathbf{N})$ is not a subset of the finite set $I_k$, and so the set $F(\mathbf{N}) - I_k$ is non-empty and has a least element, $F(t)$, for some $t \in \mathbf{N}$. Since $F(1) = l(A) \leqslant k$, $t \neq 1$. Hence $t = S(q)$ $(q \in \mathbf{N})$, and we have $F(q) \leqslant k < F(S(q)) = F(t)$. Since, by the definition of $F$, $F(S(q)) = l(A - I_{F(q)})$, it follows that $F(q) = k$. But then $F$ is surjective, and is therefore a bijection from **N** to $A$. Thus, $A$ is countable.

*Exercise 6.1* If C is a non-empty set of sets, then equipotence acts as an equivalence relation in C.

*Exercise 6.2* If $A$, $B$ are non-empty sets, then $A \precsim B$ if and only if $A \sim B$ or $A < B$.

*Exercise 6.3* Let $A$, $B$ be sets such that $A \sim B$. Prove:
If $A$ is finite, then $B$ is finite.
If $A$ is countable, then $B$ is countable.

*Exercise 6.4* If $n < m$ $(n, m \in \mathbf{N})$, then $I_n \prec I_m$.

*Exercise 6.5* If $n \in \mathbf{N}$, then a set $Y$ "has $n$ elements" if $Y \sim I_n$. Prove: if $Y \subset X$ and $t \in X - Y$, then $Y$ has $n$ elements if and only if $Y \cup \{t\}$ has $n + 1$ elements.

*Exercise 6.6*
(a) **N** is infinite.
(b) Every countable set is infinite.

*Exercise 6.7* A subset $K$ of **N** is an initial segment if and only if

(a) $S(n) \in K \Rightarrow n \in K$

and

(b) there is some $m \in K$ such that $S(m) \notin K$.

*Exercise 6.8* For each $n \in \mathbf{N}$, the terminal segment, $T_n$, is the set $T_n = \{m \in N | m \geqslant n\}$. Prove:

(a) Every terminal segment is infinite.

(b) A non-empty subset of **N** is a terminal segment if and only if it is an inductive set.

*Exercise 6.9*

(a) Every initial segment is finite.

(b) No terminal segment is finite.

*Exercise 6.10* Every subset of a finite set is finite.

Hint: use induction!

*Exercise 6.11*

(a) If $A$ and $B$ are finite sets, then $A \cup B$ and $A \cap B$ are finite.

(b) If $K$ is a finite set of finite sets, then $\bigcup_{A \in K} A$ and $\bigcap_{A \in K} A$ are finite sets.

(c) Does (b) generalize to arbitrary sets, $K$, of finite sets?

• *Exercise 6.12* Let $A$ be a set, $<$ an order relation on $A$. An element $m \in A$ such that $m \leqslant a$ for all $a \in A$ is a *minimum* element of $A$; an element $M \in A$ such that $a \leqslant M$ for all $a \in A$ is a *maximum* element of $A$. Prove that every finite ordered set has a unique maximum and a unique minimum.

*Exercise 6.13* Every countable set is infinite.

*Exercise 6.14* Every infinite set has a countable subset.

*Exercise 6.15* Every subset of a countable set is either finite or countable.

*Exercise 6.16*

(a) If $F: A \rightarrow B$, then $B \precsim A$.

(b) If $F$ is a function with countable domain, then the range of $F$ is either finite or countable.

*Exercise 6.17* If $K$ is a countable set of non-empty finite sets, then $\bigcup_{A \in K} A$ is countable.

*Exercise 6.18* If $K$ is a countable set of countable sets, then $\bigcup_{A \in K} A$ is countable.

*Exercise 6.19* If $A$ is a countable set, then so is $A \times A$.

*Exercise 6.20* If $A \sim I_n$ $(n \in N)$, then $\mathscr{P}(A) \sim I_{n!}$, where $n! = \prod_{i=1}^{n} i$.

*Exercise 6.21*

(a) If $A$ is finite and $B$ is infinite, then $A \prec B$.

(b) If $A$ is countable and $B$ is infinite, then $A \precsim B$.

7. **SEQUENCES, TUPLES, FAMILIES.** In §2, we defined a sequence, $(a_n)$, in a set $A$ as a function from **N** to $A$. (Definition 2.1)

Continuing in the same spirit, we define, for each $n \in \mathbf{N}$, an *n-tuple in a set* $A$ as a function from $I_n$ to $A$. We use the symbol $\langle a_1, \ldots, a_n \rangle$ to designate the $n$-tuple $F: I_n \to A$ such that $F(i) = a_i$ for each $i \in I_n$.

More generally, if $I$ is any set, we may use it as an "index set," much as we used **N** or $I_n$ $(n \in \mathbf{N})$. If $A$ is any set, then a *family in A, indexed on I,* is a function $F: I \to A$. We write $\{a_i\}_{i \in I}$ to designate the family $F: I \to A$ such that $F(i) = a_i$ for each $i \in I$.

*Exercise 7.1* Let $A$ be a non-empty set and let $\overline{A}$ be the set of all 2-tuples $\langle x, y \rangle$, where $x, y \in A$. Prove: if $\langle x, y \rangle, \langle z, t \rangle \in \overline{A}$, then $\langle x, y \rangle = \langle z, t \rangle$ if and only if $x = z$ and $y = t$. (Thus, 2-tuples behave exactly like the ordered pairs we introduced in Chapter 0, and we generally identify $(x, y)$ and $\langle x, y \rangle$.)

*Exercise 7.2* Prove that the set of all $n$-tuples $(n \in \mathbf{N})$ in an infinite set $A$ is equipotent to $A$.

# CHAPTER 2

# THE INTEGERS

**1. THE SET OF INTEGERS.** The system $\langle \mathbf{N}, +, \cdot, < \rangle$, where $\mathbf{N}$ is the set of all natural numbers, reflects the properties of the familiar whole numbers with respect to addition, multiplication, and order. Since $m + p \neq m$ for all $m, p \in \mathbf{N}$, there are no natural numbers corresponding to zero or to the negative whole numbers. We will construct from $\mathbf{N}$ a set $\mathbf{Z}$ whose elements we call integers and will define in $\mathbf{Z}$ an addition $(+_{\mathbf{Z}})$, a multiplication $(\cdot_{\mathbf{Z}})$ and an order $(<_{\mathbf{Z}})$ in such a way that $\mathbf{Z}$ will reflect the properties of the familiar positive, zero, and negative whole numbers. The resulting system $\langle \mathbf{Z}, +_{\mathbf{Z}}, \cdot_{\mathbf{Z}}, <_{\mathbf{Z}} \rangle$ will be an extension of $\langle \mathbf{N}, +, \cdot, < \rangle$ in the sense that there exists a 1–1 mapping of $\mathbf{N}$ into $\mathbf{Z}$ which "preserves" addition, multiplication, and order. The integers will correspond to the "signed whole numbers."

We observe that every signed whole number can be represented in a variety of ways as a difference of two whole numbers (e.g., $+3 = 4 - 1 = 10 - 7 = 12 - 9$; $-2 = 1 - 3 = 5 - 7 = 18 - 20$), and that two such differences are equal when their "cross-sums" are equal (e.g., $4 + 7 = 10 + 1$; $1 + 7 = 5 + 3$). To reflect these properties of the signed whole numbers, we define an integer as an equivalence class of ordered pairs of natural numbers (corresponding to the differences of whole numbers), such that $(m, n)$ and $(p, q)$ are equivalent if the "cross-sums" $m + q$ and $p + n$ are equal.

THEOREM 1.1 *There is an equivalence relation $Q$ in $\mathbf{N} \times \mathbf{N}$ such that $(m, n)\, Q\, (p, q)$ holds whenever $m + q = p + n$ in $\mathbf{N}$.*

PROOF: Since the set $Q = \{((m, n), (p, q)) \mid m + q = p + n; m, n, p, q \in \mathbf{N}\}$ is a subset of $(\mathbf{N} \times \mathbf{N}) \times (\mathbf{N} \times \mathbf{N})$, $Q$ is a binary relation in $\mathbf{N} \times \mathbf{N}$ (Definition 0, 3.1). Since $m + n = m + n$, $((m, n), (m, n)) \in Q$, and $Q$ is reflexive. If $((m, n), (p, q)) \in Q$, then $m + q = p + n$. Hence, $p + n = m + q$, so that $((p, q), (m, n)) \in Q$ and $Q$ is symmetric. Finally, if $((m, n), (p, q))$ and $((p, q), (r, s)) \in Q$, then $m + q = p + n$ and $p + s = r + q$. But then

$$(m+q)+s=(p+n)+s=(p+s)+n=(r+q)+n.$$

Hence, $q+(m+s)=q+(r+n)$ and, by the cancellation law in **N**, $m+s=r+n$, so that $((m, n), (r, s)) \in Q$. Thus, $Q$ is transitive. By Definition 0, 3.3 $Q$ is an equivalence relation.

DEFINITION 1.1 We write "$(m, n) \sim (p, q)$" for "$((m, n), (p, q)) \in Q$" $\in Q$" and read "$\sim$" as "is equivalent to." For each $(m, n) \in \mathbf{N} \times \mathbf{N}$, $C_{(m,n)}$ is the set of all $(p, q) \in \mathbf{N} \times \mathbf{N}$ such that $(p, q) \sim (m, n)$. An *integer* is an equivalence class $C_{(m,n)}$. We write "**Z**" for the set of all integers and "$a$," "$b$," "$c$," ... for elements of **Z**.

The set **Z** of all integers is the factor set $(\mathbf{N} \times \mathbf{N})/Q$, where $Q$ is the subset of $(\mathbf{N} \times \mathbf{N}) \times (\mathbf{N} \times \mathbf{N})$ defined in Theorem 1.1 (Definition 0, 3.3).

**2. ADDITION IN Z.** The ordered pairs of natural numbers which constitute an integer correspond to the differences of whole numbers associated with a signed whole number. Termwise addition of differences associated with two signed numbers gives a difference associated with their sum—for example.

$$+3+(-2)=(4-1)+(1-3)=(4+1)-(1+3).$$

This suggests that *componentwise* addition of ordered pairs belonging to two integers should give an ordered pair belonging to the sum of the integers, i.e. that the sum of $a = C_{(m,n)}$ and $b = C_{(p,q)}$ should be the integer $c = C_{(m+p,n+q)}$. We shall show that $c$ is independent of the choice of $(m, n) \in a$ and $(p, q) \in b$.

THEOREM 2.1 *If $(m', n') \sim (m, n)$ and $(p', q') \sim (p, q)$, then*

$$(m+p, n+q) \sim (m'+p', n'+q').$$

PROOF: By hypothesis and Definition 1.1, $m'+n=m+n'$ and $p'+q=p+q'$. Hence, by the properties of addition in **N**,

$$\begin{aligned}(m+p)+(n'+q') &= (m+n')+(p+q')\\ &= (m'+n)+(p'+q)=(m'+p')+(n+q),\end{aligned}$$

and, by Definition 1.1, $(m+p, n+q) \sim (m'+p', n'+q')$.

THEOREM 2.2 *There is a binary operation $F$ in **Z** such that*

$$F(a, b) = C_{(m+p,n+q)}$$

*if $(m, n) \in a$ and $(p, q) \in b$.*

PROOF: The set

$$F=\{((a, b), C_{(m+p,n+q)} \mid (m, n)\in a; (p, q)\in b; a, b\in \mathbf{Z}\}$$

is a subset of $(\mathbf{Z} \times \mathbf{Z}) \times \mathbf{Z}$. For each $(a, b) \in \mathbf{Z} \times \mathbf{Z}$ there are $(m, n) \in a$, $(p, q) \in b$, and $c = C_{(m+p,n+q)}$ such that $((a, b), c) \in F$. If $(m', n') \in a$, $(p', q') \in b$, and $c' = C_{(m'+p',n'+q')}$ then $(m', n') \sim (m, n)$, $(p', q') \sim (p, q)$ and, by Theorem 2.2, $(m + p, n + q) \sim (m' + p', n' + q')$. Hence, by Definition 2.1, $c' = c$, and $F$ is a mapping of $\mathbf{Z} \times \mathbf{Z}$ into $\mathbf{Z}$, by Definition 0.13. By Definition 0.16, $F$ is a binary operation in $\mathbf{Z}$ and $F(a, b) = c$.

DEFINITION 2.1 We call the binary operation $F$ of Theorem 2.2 *addition* in $\mathbf{Z}$ and write $a +_{\mathbf{Z}} b = F(a, b)$ for all $a, b \in \mathbf{Z}$. (We omit the subscript and write $a + b$ if no confusion arises in the context.)

THEOREM 2.3 *Addition in* $\mathbf{Z}$ *is associative and commutative.*

PROOF: (1) Addition in $\mathbf{Z}$ is associative.
If $a, b, c\in \mathbf{Z}$, then $a = C_{(m,n)}, b = C_{(p,q)}, c = C_{(r,t)}$ for some $m, n, p, q, r, t \in \mathbf{N}$. By Definition 2.2, Theorem 2.2, and the associativity of addition in $\mathbf{N}$,

$$\begin{aligned} a +_{\mathbf{Z}} (b +_{\mathbf{Z}} c) &= C_{(m,n)} +_{\mathbf{Z}} C_{(p+r,q+t)} \\ &= C_{(m+p,n+q)} +_{\mathbf{Z}} C_{(r,t)} \\ &= (a +_{\mathbf{Z}} b) +_{\mathbf{Z}} c. \end{aligned}$$

(2) Addition in $\mathbf{Z}$ is commutative.
If $a, b \in \mathbf{Z}$, then $a = C_{(m,n)}, b = C_{(p,q)}$ for some $m, n, p, q \in \mathbf{N}$. By Definition 2.2, Theorem 2.2, and the commutativity of addition in $\mathbf{N}$,

$$\begin{aligned} a +_{\mathbf{Z}} b = F(a, b) &= C_{(m+p,n+q)} \\ &= C_{(p+m,q+n)} = F(b, a) = b +_{\mathbf{Z}} a. \end{aligned}$$

THEOREM 2.4 *For each* $a \in \mathbf{Z}$, $a +_{\mathbf{Z}} C_{(1,1)} = a = C_{(1,1)} +_{\mathbf{Z}} a$.

PROOF: If $a = C_{(m,n)}$, then $C_{(1,1)} +_{\mathbf{Z}} a = a +_{\mathbf{Z}} C_{(1,1)} = C_{(m,n)} + C_{(1,1)} = C_{(m+1,n+1)} = C_{(m,n)} = a$.

*Notation:* We write "$0_{\mathbf{Z}}$" for the element $C_{(1,1)}$.

THEOREM 2.5 *If* $a \in \mathbf{Z}$, *then there is exactly one element* $a' \in \mathbf{Z}$ *such that* $a +_{\mathbf{Z}} a' = a' +_{\mathbf{Z}} a = 0$.

PROOF: If $a = C_{(m,n)} \in \mathbf{Z}$ and $a' = C_{(n,m)}$ then

$$a +_{\mathbf{Z}} a' = C_{(m+n,n+m)}.$$

Since $(q, q) \sim (1, 1)$ for all $q \in \mathbf{N}$, it follows from Definition 1.1 that $C_{(m+n,m+n)} = C_{(1,1)}$. Hence $a +_{Z} a' = a' +_{Z} a = 0$, and $a'$ has the required properties.

If $a''$ is another element of $\mathbf{Z}$ such that

$$a'' + a = a + a'' = 0,$$

then $a' = a' + 0 = a' + (a + a'') = (a' + a) + a'' = 0 + a'' = a''$.

*Notation:* We write "$-a$" for the element $a'$ of Theorem 2.5.

*Exercise 2.1*

(a) For $a, b \in \mathbf{Z}$ there exists a unique $c \in \mathbf{Z}$ such that $a = b + c$ in $\mathbf{Z}$.
(b) If $a + c = b + c$ in $\mathbf{Z}$, then $a = b$ (cancellation law for addition in $\mathbf{Z}$).

*Notation:* We write "$a - b$" for the unique element $c$ such that $a = b + c$ in $\mathbf{Z}$.

## 3. IDENTITIES; INVERSES; GROUPS

DEFINITION 3.1 Let $A$ be a set, $e \in A$, and $\circ$ a binary operation on $A$. Then $e$ is an *identity for* $\circ$ if $a \circ e = a = e \circ a$ for each $a \in A$.

DEFINITION 3.2 Let $A$ be a set, $\circ$ a binary operation on $A$, $e \in A$ an identity for $\circ$. If $a, a' \in A$, then $a'$ is an *inverse of a with respect to the operation* $\circ$ *and the identity e* if $a \circ a' = e = a' \circ a$.

THEOREM 3.1 *If* $\circ$ *is a binary operation on a set A, then there is at most one identity for* $\circ$ *in A.*

PROOF: If $e, e' \in A$ are both identities for $\circ$, then $e = e \circ e' = e'$

THEOREM 3.2 *Let A be a set,* $\circ$ *an associative binary operation on A and* $e \in A$ *an identity for* $\circ$*. Then an element* $a \in A$ *has at most one inverse in A with respect to the operation* $\circ$ *and the identity e.*

PROOF: Suppose $a', a'' \in A$ are both inverses of $A$. Then

$$(a' \circ a) \circ a'' = a' \circ (a \circ a'')$$
$$e \circ a'' = a' \circ e$$
$$a'' = a'.$$

DEFINITION 3.3 Let $A$ be a set, $\circ$ a binary operation on $A$. Then $\langle A, \circ \rangle$ is a *group* if

(1) $\circ$ is associative;
(2) $A$ contains an identity, $e$, for $\circ$;
(3) every element of $A$ has an inverse with respect to the operation $\circ$ and the identity $e$.

The group $\langle A, \circ \rangle$ is *abelian* if the operation $\circ$ is commutative.

THEOREM 3.3 *$\langle \mathbf{Z}, +_{\mathbf{Z}} \rangle$ is an abelian group.*

PROOF: By Theorem 2.3, $+_{\mathbf{Z}}$ is associative and commutative. By Theorems 2.4 and 3.2, $0_{\mathbf{Z}} = C_{(1,1)}$ is the identity for $+_{\mathbf{Z}}$. By Theorem 2.5, every integer $a = C_{(m,n)}$ has additive inverse $-a = C_{(n,m)}$ with respect to $0_{\mathbf{Z}}$.

*Exercise 3.1* The cancellation laws hold in a group, i.e., if $\langle A, \circ \rangle$ is a group, then for $a, b, x \in A$,

$$a \circ x = b \circ x \text{ implies } a = b$$

and

$$x \circ a = x \circ b \text{ implies } a = b.$$

*Exercise 3.2* Let $A$ be a set, $\circ$ an associative binary operation on $A$. Suppose

(1) there is an element $e \in A$ such that $a \circ e = a$ for each $a \in A$,

and

(2) for each $a \in A$, there is an element $a' \in A$ such that $a \circ a' = e$.

Prove that $\langle A, \circ \rangle$ is a group.

*Exercise 3.3* Let $\circ$ be an associative binary operation on a set $A$. Prove that $\langle A, \circ \rangle$ is a group if and only if, for each $a, b \in A$, the equations $a \circ x = b$ and $y \circ a = b$ have solutions in $A$.

Hint: Prove that $\langle A, \circ \rangle$ satisfies the conditions of Exercise 3.2.

*Exercise 3.4* Let $\circ$ be an associative binary operation on a *finite* set $A$. Suppose that both cancellation laws hold in $A$, i.e.,

(1) $a \circ x = b \circ x \Rightarrow a = b$

and

(2) $x \circ a = x \circ b \Rightarrow a = b$.

(a) Prove that $\langle A, \circ \rangle$ is a group.

(b) Give an example of an associative binary operation on a *infinite* set $A$ such that (1) and (2) hold and $\langle A, \circ \rangle$ is *not* a group.

Hint for (a): Prove that, if $A = \{x_1, \ldots, x_n\}$, $a \in A$, then $A = \{ax_1, \ldots, ax_n\}$. Now use the result of Exercise 3.4.

*Exercise 3.5* For any non-empty set $A$, let $S_A$ be the set of all bijections from $A$ to $A$.

(1) Prove that $\langle S_A, \circ\rangle$ is a group, where $\circ$ is the operation "composition of functions" (i.e., for $F, G \in S_A$, $(F \circ G)(a) = F(G(a))$ for each $a \in A$).
(2) Prove that $S_A$ is non-abelian except when $A$ consists of a single element, or of two elements.
(3) If $A$ is a finite set having $n$ elements, prove that $S_A$ has $n! = \prod_{i=1}^{n} i$ elements.

*Exercise 3.6* Let $\langle A, +\rangle$ be an abelian group, and let "$-a$" and "$a - b$" denote, respectively, the inverse of $a$ and the element $c \in A$ such that $a = b + c$. Then, for each $a, b, c \in A$,

(1) $-(-a) = a$.
(2) $a + (-b) = a - b$.
(3) $-(a + b) = (-a) + (-b) = -a - b$.
(4) $(a - b) + (b - c) = a - c$.
(5) $-(a - b) = b - a$.

**4. MULTIPLICATION IN Z.** Our definition of multiplication in **Z** is patterned on the behavior of differences of signed whole numbers under multiplication:

$$(+2)(-3) = (4 - 2)(3 - 6) = (4 \cdot 3 + 2 \cdot 6) - (2 \cdot 3 + 4 \cdot 6).$$

This suggests that the product of $a = C_{(m,n)}$ and $b = C_{(p,q)}$ should be $c = C_{(mp+nq,mq+np)}$, provided that $c$ is independent of the choice of $(m, n) \in a$ and $(p, q) \in b$.

THEOREM 4.1 *If $(m, n) \sim (m', n')$ and $(p, q) \sim (p', q')$, then*

$$(mp + nq, mq + np) \sim (m'p' + n'q', m'q' + n'p').$$

PROOF: The conclusion of the theorem follows from

(1) $(mp + nq, mq + np) \sim (m'p + n'q, m'q + n'p)$

and

(2) $(m'p + n'q, m'q + n'p) \sim (m'p' + n'q', m'q' + n'p')$

by the transitivity of the equivalence relation.

By hypothesis, $m + n' = m' + n$. Hence, by the properties of addition and multiplication in **N**,

$$\begin{aligned}(mp + nq) + (m'q + n'p) &= (m + n')p + (n + m')q \\ &= (m + n')(p + q)\end{aligned}$$

and

$$(m'p + n'q) + (mq + np) = (m' + n)p + (n' + m)q$$
$$= (m + n')(p + q).$$

Thus, by Definition 1.1, (1) is proved.

By hypothesis, $p + q' = p' + q$. Hence, by the properties of addition and multiplication in **N**,

$$(m'p + n'q) + (m'q' + n'p') = m'(p + q') + n'(q + p')$$
$$= (m' + n')(p + q')$$

and

$$(m'p + n'q') + (m'q + n'p) = m'(p' + q) + n'(p + q')$$
$$= (m' + n')(p + q')$$

Thus, by Definition 1.1, (2) is proved.

THEOREM 4.2 *There is a binary operation G on* **Z** *such that*

$$G(a, b) = C_{(mp+nq, mq+np)}$$

*if* $(m, n) \in a$ *and* $(p, q) \in b$.

PROOF: The set

$$G = \{((a, b), C_{(mp+nq, mq+np)}) | (m, n) \in a; (p, q) \in b; a, b \in \mathbf{Z}\}$$

is a subset of $(\mathbf{Z} \times \mathbf{Z}) \times \mathbf{Z}$. For each $(a, b) \in \mathbf{Z} \times \mathbf{Z}$ there are $(m, n) \in a$, $(p, q) \in b$, and $c = C_{(mp+nq, mq+np)}$ such that $((a, b), c) \in G$. If $(m', n') \in a$, $(p', q') \in b$, and $c' = C_{(m'p'+n'q', m'q'+n'q')}$, then $(m', n') \sim (m, n)$, $(p', q') \sim (p, q)$ and, by Theorem 2.7, $(m'p' + n'q', m'q' + n'p') \sim (mp + nq, mq + np)$.

But then there is a function $G\colon \mathbf{Z} \times \mathbf{Z} \to \mathbf{Z}$ such that

$$G(a, b) = C_{(mp+nq, mq+np)}$$

if $(m, n) \in a$, $(p, q) \in b$. By Definition 1, 3.1, $G$ is a binary operation on **Z**.

DEFINITION 4.1 We call the binary operation $G$ of Theorem 4.2 *multiplication* in **Z** and write

$$a \cdot_{\mathbf{Z}} b = G(a, b) \text{ for all } a, b \in \mathbf{Z}.$$

(We omit the subscript and write $a \cdot b$ or $ab$ if no confusion arises in the context.)

THEOREM 4.3 *Multiplication in* $\mathbf{Z}$ *is associative, commutative and distributive over addition. The integer* $C_{(1+1,1)}$ *serves as the identity for multiplication.*

*We leave the proof as an exercise. (Exercise 4.2)*

*Notation:* We write "$1_Z$" for the identity for multiplication. (If no confusion with "1" in $\mathbf{N}$ is likely, we omit the subscript and write "1.")

• *Exercise 4.1* Prove that $0_Z \neq 1_Z$.

• *Exercise 4.2* Prove Theorem 4.3.

• *Exercise 4.3* Prove that $1_Z$ and $-1_Z$ are the only integers which have multiplication inverses in $\mathbf{Z}$.

*Exercise 4.4* Prove that $\langle \mathbf{Z}, \cdot_Z \rangle$ is not a group.

## 5. RINGS.

DEFINITION 5.1 Let $A$ be a set, $+$, $\cdot$ binary operations on $A$. Then $\langle A, +, \cdot \rangle$ is a *ring* if

(1) $\langle A, + \rangle$ is an abelian group;

(2) $\cdot$ is associative;

(3) $\cdot$ is right and left distributive over $+$.

(We shall refer to $+$ and $\cdot$ as *addition* and *multiplication,* respectively. If no confusion is likely, we shall refer to $\langle A, +, \cdot \rangle$ as "the ring $A$.")

A ring $A$ is called a *commutative* ring if its multiplication is commutative, and is called a ring *with identity* if $A$ contains an identity for multiplication.

*Notation:* We write "0" for the additive identity of a ring $A$, $-a$ for the additive inverse of $a \in A$. If $A$ is a ring with identity, we write 1 to designate the multiplicative identity of $A$.

THEOREM 5.1 $\langle \mathbf{Z}, +_{\mathbf{Z}}, \cdot_Z \rangle$ is a commutative ring with identity.

PROOF: This follows immediately from Theorems 3.3 and 4.3.

If $\langle A, +, \cdot \rangle$ is a ring, we have the following calculation rules, besides those stated for $\langle A, + \rangle$ in Exercise 3.6:

THEOREM 5.2 *For* $a, b, c \in A$,

(1) $a \cdot 0_A = 0_A \cdot a = 0_A$,

(2) $(-a) \cdot b = a \cdot (-b) = -a \cdot b$,

(3) $(-a) \cdot (-b) = a \cdot b$,

(4) $\begin{cases} (-a) \cdot (b + c) = -a \cdot b - a \cdot c, \\ (b + c) \cdot (-a) = -b \cdot a - c \cdot a. \end{cases}$

*We leave the proof as an exercise (Exercise 5.2).*

*Exercise 5.1* If $M$ is the set of all $2 \times 2$ matrices of integers, where addition in $M$ is defined by

$$\begin{pmatrix} a & b \\ c & d \end{pmatrix} + \begin{pmatrix} e & f \\ g & h \end{pmatrix} = \begin{pmatrix} a+e & b+f \\ c+g & d+h \end{pmatrix}$$

and multiplication in $M$ is defined by

$$\begin{pmatrix} a & b \\ c & d \end{pmatrix} \cdot \begin{pmatrix} e & f \\ g & h \end{pmatrix} = \begin{pmatrix} ae+bg & af+bh \\ ce+dg & cf+dh \end{pmatrix},$$

then $\langle M, +, \cdot \rangle$ is a non-commutative ring with identity.

*Exercise 5.2* Prove Theorem 5.2.

**6. ORDER IN Z.** The familiar integers are positive, negative, or zero. An integer $b$ exceeds an integer $a$ if the difference $b - a$ is positive. This suggests introducing in **Z** a set of positive elements and using it to define order.

DEFINITION 6.1 An integer $a$ is *positive* if $n < m$ for some $(m, n) \in a$.

It is useful to have several equivalent characterizations of positivity in **Z**.

THEOREM 6.1 *The following conditions on an integer a are equivalent:*

(1) *$n < m$ for some $(m, n) \in a$ (i.e., $a$ is positive according to Definition 6.1);*

(2) *$p < q$ for all $(p, q) \in a$;*

(3) $a = C_{(t+1,1)}$ for some $t \in \mathbf{N}$.

PROOF: $1 \Rightarrow 2$ Suppose $n < m$ for some $(m, n) \in a$. Let $(p, q) \in a$. Then $(m, n) \sim (p, q)$, hence $m + q = p + n$. Since $n < m$, $m = n + h$ for some $h \in \mathbf{N}$. Hence $(n + h) + q = p + n$ and, by the properties of addition in **N**, it follows that $p = q + h$, and $q < p$.

$2 \Rightarrow 3$ Let $(p, q) \in a$. Then $1 \leqslant q < p$, whence $p = r + 1$ for some $r \in \mathbf{N}$. If $q = 1$, we have $(r + 1, 1) \in a$, hence $a = C_{(r+1,1)}$ is of the required form. Suppose $1 < q < p$. Then $q = s + 1$, $s \in \mathbf{N}$, and from $s + 1 = q < p = r + 1$, we conclude that $s < r$, hence $r = s + t$ $(t \in \mathbf{N})$. But then $a = C_{(p,q)} = C_{(r+1,s+1)} = C_{(s+t+1,s+1)} = C_{(t+1,1)}$, as required.

$3 \Rightarrow 1$ If $a = C_{(t+1,1)}$, $t \in \mathbf{N}$, then $1 < t + 1$ and $(t + 1, 1) \in a$.

*Notation:* We denote by $\mathbf{Z}^+$ the set of all positive integers.

THEOREM 6.2 *If $a \in \mathbf{Z}$, then one and only one of the following holds:*

$$(1)\ a \in \mathbf{Z}^+ \quad (2)\ a = 0_{\mathbf{Z}} \quad (3)\ -a \in \mathbf{Z}^+.$$

PROOF: Let $a = C_{(m,n)}$. By Theorem 6.1, $a \in \mathbf{Z}^+$ if and only if $n < m$. Since $C_{(1,1)} = 0$, $a = 0$ if and only if $m = n$. Since $C_{(n,m)} = -C_{(m,n)}$, $-a \in \mathbf{Z}^+$ if and only if $m < n$. By trichotomy of order in $\mathbf{N}$, exactly one of $m < n, m = n, n < m$ holds. Hence exactly one of (1), (2), (3) holds.

THEOREM 6.3 *If $a, b \in \mathbf{Z}^+$, then $a + b \in \mathbf{Z}^+$ and $ab \in \mathbf{Z}^+$.*

PROOF: If $a, b \in \mathbf{Z}^+$, then $a = C_{(m,n)}, b = C_{(p,q)}$, where $n < m$ and $q < p$. Hence $n + q < m + p$, and $a + b = C_{(m+p,n+q)} \in \mathbf{Z}^+$.

To prove that, if $a, b \in \mathbf{Z}^+$, then $ab \in \mathbf{Z}^+$, we use the third characterization of positivity given in Theorem 6.1. Accordingly, we have $a = C_{(s+1,1)}$, $b = C_{(t+1,1)}$, for some $s, t \in \mathbf{N}$. But then $ab = C_{((s+1)(t+1)+1,s+1+t+1)} = C_{(st+s+t+1+1,s+t+1+1)} = C_{(st+1,1)} \in \mathbf{Z}^+$.

THEOREM 6.4 *If $T$ is the set of all elements $(a, b) \in \mathbf{Z} \times \mathbf{Z}$ such that $b - a$ is positive, then $T$ is an order relation in $\mathbf{Z}$.*

PROOF: If $a, b \in \mathbf{Z}$, then $b - a \in \mathbf{Z}$ and, by Theorem 6.2, just one of

$$a - b = 0, \qquad a - b \text{ is positive}, \qquad b - a \text{ is positive}$$

holds. Hence, just one of

$$a = b, \qquad (a, b) \in T, \qquad (b, a) \in T$$

holds, and $T$ has the property of trichotomy.

If $(a, b) \in T$ and $(b, c) \in T$, then $b - a$ and $c - b$ are positive and, by Theorem 6.3, $c - a = (c - b) + (b - a)$ is positive and $(a, c) \in T$. Hence, $T$ is transitive. By Definition 0, 3.6, $T$ is an order relation in $\mathbf{Z}$.

*Notation:* We write "$a <_{\mathbf{Z}} b$" or "$b <_{\mathbf{Z}} a$" (read "$a$ is less than $b$ in $\mathbf{Z}$" or "$b$ is greater than $a$ in $\mathbf{Z}$") if $b - a$ is positive, i.e., if $(a, b) \in T$, where $T$ is the order relation of Theorem 6.4. Usually we omit the subscript and write "$a < b$" or "$b > a$."

Addition of an element, or multiplication by a positive element, do not disturb inequalities in $\mathbf{Z}$, nor does cancellation of terms in sums, or of positive factors in products. More precisely, we have

THEOREM 6.5 *For $a$, $b$ in* $\mathbf{Z}$, *$a < b$ if and only if* (1) $a + c < b + c$ *for all $c \in \mathbf{Z}$, or* (2) $ac < bc$ *for all* $c \in \mathbf{Z}^+$.

PROOF: See Exercise 6.3.

● *Exercise 6.1* $\mathbf{Z}^+ = \{a \in \mathbf{Z} \,|\, a > 0\} = \{-a \in \mathbf{Z} \,|\, a < 0\}$

● *Exercise 6.2* If $a \neq 0, b \neq 0$ in $\mathbf{Z}$, then $a \cdot b \neq 0$.

● *Exercise 6.3* Prove Theorem 6.5.

*Exercise 6.4* If in a ring $\langle A, +, \cdot\rangle$ any one of the following statements holds, then the other two hold also:

(1) If $a$, $b \in A$ and $a \cdot b = 0_A$, then $a = 0_A$ or $b = 0_A$. ($A$ has no non-zero "divisors of zero.")

(2) If $a$, $b \in A$, $c \neq 0_A$, and $a \cdot c = b \cdot c$, then $a = b$. (Right-cancellation.)

(3) If $a$, $b \in A$, $c \neq 0_A$, and $c \cdot a - c \cdot b$, then $a = b$. (Left-cancellation.)

## 7. ORDERED INTEGRAL DOMAINS.

In this section, we place the ideas introduced in the preceding section into a more general setting. This will be useful to us later in defining order for rational numbers and for real numbers.

DEFINITION 7.1 If $\langle A, +, \cdot\rangle$ is a commutative ring with identity $1_A \neq 0_A$ such that, for $a$, $b \in A$, $a \cdot b = 0_A$ only if $a = 0_A$ or $b = 0_A$, then $\langle A, +, \cdot\rangle$ is called an *integral domain.*

THEOREM 7.1 $\langle \mathbf{Z}, +, \cdot\rangle$ *is an integral domain.*

PROOF: By Theorem 5.1 and Exercise 4.1, $\mathbf{Z}$ is a commutative ring with identity $1_\mathbf{Z} \neq 0_\mathbf{Z}$. By Exercise 6.2, the product of two non-zero integers is different from zero. Hence $\mathbf{Z}$ is an integral domain.

DEFINITION 7.2 If $\langle A, +, \cdot\rangle$ is an integral domain and $<$ is an order relation in $A$ such that

(1) for $a < b$ in $A$, $a + c < b + c$ for all $c \in A$, and

(2) for $a < b$ in $A$, $a \cdot c < b \cdot c$ for all $c > 0_A$ in $A$, then the system $\langle A, +, \cdot, <\rangle$ is called an *ordered integral domain.*

COROLLARY: $\langle \mathbf{Z}, +_\mathbf{Z}, \cdot_\mathbf{Z}, <_\mathbf{Z}\rangle$ *is an ordered integral domain.*

DEFINITION 7.3 If $\langle A, +, \cdot\rangle$ is an integral domain, then a subset $A^+$ of $A$ is called a *set of positive elements for A* if

(1) $a + b \in A^+$ for all $a$, $b \in A^+$,

(2) $a \cdot b \in A^+$ for all $a$, $b \in A^+$,

(3) for $a \in A$, exactly one of the following holds:

$$a \in A^+, \qquad a = 0_A, \qquad -a \in A^+.$$

COROLLARY: *$\mathbf{Z}^+$ is a set of positive elements for $\mathbf{Z}$.*

THEOREM 7.2 *If $\langle A, +, \cdot\rangle$ is an integral domain and $A^+$ is a set of positive elements for $A$, then*

(1) *the subset $T$ of $A \times A$ defined by*

$$T = \{(a, b) | b - a \in A^+\}$$

*is an order relation in $A$.*

(2) *If we write "$a < b$" ("$b > a$") for "$(a, b) \in T$," then $\langle A, +, \cdot, <\rangle$ is an ordered integral domain.*

(3) $A^+ = \{a | a > 0_A\}$

PROOF: If $(a, b)$ and $(b, c)$ are in $T$, then $b \cdot -a$ and $c - b$ are in $A^+$. Hence, by Definition 7.3 (2),

$$c - a = (c - b) + (b - a) \in A^+$$

Thus, $(a, c) \in T$, and $T$ is transitive.

For $a, b \in A$, exactly one of

$$b - a \in A^+, \qquad b - a = 0, \qquad -(b - a) = a - b \in A^+$$

holds, by Definition 7.3 (3). Thus, $T$ is trichotomous. It follows that $T$ is an order relation in $A$.

If $(a, b) \in T$, then, for any $c \in A$,

$$b + c - (a + c) = b - a \in A^+.$$

Hence, $(a + c, b + c) \in T$. Further, if $(a, b) \in T$ and $c \in A^+$ then

$$bc - ac = (b - a)c \in A^+$$

by Definition 7.3 (1). Hence, $(ac, bc) \in T$.

Thus, from $a < b$ and $c \in A$, follows $a + c < b + c$, and from $a < b$ and $c \in A^+$ follows $ac < bc$, so that $\langle A, +, \cdot, <\rangle$ is an ordered integral domain.

*Exercise 7.1*

(a) Any commutative ring with identity $1 \neq 0$ in which is defined an order relation satisfying (1) and (2) of Definition 7.2 is an ordered integral domain.

(b) Any commutative ring with identity $1 \neq 0$ which contains a set of positive elements, in the sense of Definition 7.3, is an ordered integral domain.

*Exercise 7.2* If $\langle A, +, \cdot, < \rangle$ is an ordered integral domain, then the set

$$A^+ = \{a \in A \mid a > 0_A\}$$

is a set of positive elements for $A$ and the order relation

$$T = \{(a, b) \in A \times A \mid b - a \in A^+\}$$

is the given order in $\langle A, +, \cdot, < \rangle$.

• *Exercise 7.3*

(1) If $\langle A, +, \cdot, < \rangle$ is an ordered integral domain, then for $a \neq 0_A$, $a \cdot a$ is positive in $A$; the multiplicative identity $1_A$ is positive.

(2) In the ordered domain $\mathbf{Z}$ of integers, $ab = 1$ if and only if $a = b = \pm 1$.

*Exercise 7.4* If $\langle A, +, \cdot, < \rangle$ is an ordered integral domain, then $A$ is an infinite set.

• *Exercise 7.5* Let $A$ be an ordered integral domain and let $a, b, c, d$ be elements of $A$ such that $a < c$ and $b < d$. Prove that $a + b < c + d$ and $ac < bd$.

**8. EMBEDDING; ISOMORPHISM.** The elements $C_{(S(n),1)} \in \mathbf{Z}^+$ behave with respect to $+_\mathbf{Z}$, $\cdot_\mathbf{Z}$, $<_\mathbf{Z}$ exactly as the natural numbers $n$ behave with respect to $+, \cdot, <$, in the sense of the following theorem.

THEOREM 8.1 *If $E: \mathbf{N} \to \mathbf{Z}$ is defined by $E(n) = C_{(n+1,1)}$ for each $n \in \mathbf{N}$, then $E$ is an injection with range $E(\mathbf{N}) = \mathbf{Z}^+$ such that, for each $m, n \in \mathbf{N}$,*

(1) $E(m + n) = E(m) +_\mathbf{Z} E(n)$,

(2) $E(m \cdot n) = E(m) \cdot_\mathbf{Z} E(n)$,

(3) $m < n \Leftrightarrow E(m) <_\mathbf{Z} E(n)$.

PROOF: The range of $E$ is $\mathbf{Z}^+$, by Theorem 6.1. We prove 1, 2 and 3 and then conclude from 3 that $E$ is injective.

(1) For $m, n \in \mathbf{N}$, $E(m) +_\mathbf{Z} E(n) = C_{(m+1,1)} +_\mathbf{Z} C_{(n+1,1)}$
$= C_{(m+1+n+1,1+1)} = C_{(m+n+1,1)} = E_{(m+n)}$.

(2) For $m, n \in \mathbf{N}$, $E(m) \cdot_\mathbf{Z} E(n) = C_{(m+1,1)} \cdot_\mathbf{Z} C_{(n+1,1)} =$
$= C_{(mn+m+n+1+1,m+1+n+1)} = C_{(mn+1,1)} = E(mn)$.

(3) If $m < n$, then $n = m + p$ $(p \in \mathbf{N})$. Hence $E(n) = E(m) +_\mathbf{Z} E(p)$. Since $E(p) \in \mathbf{Z}^+$, we conclude that $E(m) <_\mathbf{Z} E(n)$. Conversely, if $E(m) <_\mathbf{Z} E(n)$, then $m < n$ since the alternatives $m = n$ and $n < m$ imply $E(m) = E(n)$ and $E(n) <_\mathbf{Z} E(m)$, respectively, each of which, by the trichotomy of $<_\mathbf{Z}$, contradicts our hypothesis.

Now, by (3) and the trichotomy of $<$ and $<_\mathbf{Z}$, we conclude that $E(m) = E(n)$ implies $m = n$, i.e., $E$ is injective.

Since $E: \mathbf{N} \to \mathbf{Z}$ is an injection such that $E(\mathbf{N}) = \mathbf{Z}^+$, the function $E^+: \mathbf{N} \to \mathbf{Z}^+$ such that $E^+(n) = E(n)$ for each $n \in \mathbf{N}$ is a bijection from $\mathbf{N}$ to $\mathbf{Z}^+$. (Note that $E^+$ is the co-restriction of $E$ to $\mathbf{Z}^+$, in the sense of Definition 0, 4.3.) The injection $E$, and the bijection $E^+$, both "preserve" addition, multiplication and order.

DEFINITION 8.2 Let $A, A'$ be sets, $E: A \to A'$.

(1) If $\circ, \circ'$ are binary operations on $A, A'$, respectively, then $E: A \to A'$ is an *isomorphic embedding of $A$ into $A'$ with respect to* $(\circ, \circ')$ if $E$ is injective and $E(a \circ b) = E(a) \circ' E(b)$ for each $a, b \in A$. The co-restriction of $E$ to its range, $E(A)$, is an *isomorphism of $A$ onto* $E(A)$ with respect to $(\circ, \circ')$.

(2) If $R, R'$ are binary relations on $A, A'$, respectively, then $E$ is an *isomorphic embedding of $A$ into $A'$ with respect to* $(R, R')$ if $a\, R\, b$ implies $E(a)\, R'\, E(b)$ for each $a, b \in A$. The co-restriction of $E$ to its range, $E(A)$, is an *isomorphism of $A$ onto $E(A)$, with respect to* $(R, R')$.

In the terminology of the preceding definition, Theorem 8.1 states that the function $E: \mathbf{N} \to \mathbf{Z}$ such that $E(n) = C_{(n+1,1)}$ for each $n \in \mathbf{N}$ is an isomorphic embedding of $\mathbf{N}$ into $\mathbf{Z}$ with respect to $(+, +_{\mathbf{Z}})$, $(\cdot, \cdot_{\mathbf{Z}})$ and $(<, <_{\mathbf{Z}})$. Since $E(\mathbf{N}) = \mathbf{Z}^+$, the co-restriction, $E^+$, of $E$ to $\mathbf{Z}^+$ is an isomorphism of $\mathbf{N}$ onto $\mathbf{Z}^+$, with respect to $(+, +_{\mathbf{Z}})$, $(\cdot, \cdot_{\mathbf{Z}})$ and $(<, <_{\mathbf{Z}})$.

In view of the isomorphism $E^+$ of $\mathbf{N}$ onto $\mathbf{Z}^+$ we henceforth use interchangeably the symbols $n$ and $C_{(n+1,1)}$. In particular, we "identify" 1 and $1_{\mathbf{Z}}$ (see Exercise 8.1).

DEFINITION 8.3 If there exists an isomorphic embedding of $A$ into $A'$ with respect to a pair of operations or relations $(\alpha, \alpha')$, then $A'$ is called an *extension* of $A$ with respect to $(\alpha, \alpha')$. If $A'$ is an extension of $A$ with respect to $(\alpha, \alpha')$, we say that $A$ can be *isomorphically embedded* in $A'$ with respect to $(\alpha, \alpha')$.

Thus, $\mathbf{Z}$ is an extension of $\mathbf{N}$ with respect to $(+, +_{\mathbf{Z}})$, $(\cdot, \cdot_{\mathbf{Z}})$, and $(<, <_{\mathbf{Z}})$, i.e., with respect to addition, multiplication, and order.

*Exercise 8.1* Prove that $E(1) = 1_{\mathbf{Z}}$.

*Exercise 8.2* Prove that $E$ "preserves successors," in the following sense: if $S^+: \mathbf{Z}^+ \to \mathbf{Z}^+$ is defined by: $S^+(a) = a +_{\mathbf{Z}} 1_{\mathbf{Z}}$ for each $a \in \mathbf{Z}^+$, then $E(S(n)) = S^+(E(n))$ for each $n \in \mathbf{N}$.

*Exercise 8.3* Let $S^+: \mathbf{Z}^+ \to \mathbf{Z}^+$ be defined as in Exercise 8.2. Prove that $\langle \mathbf{Z}^+, S^+ \rangle$ satisfies $A_1, A_2, A_3$ of Axiom 1.1, Chapter 1, where $\mathbf{Z}^+$ and $S^+$ play the roles of $\mathbf{N}$ and $S$, respectively.

# CHAPTER 3

# RATIONAL NUMBERS—ORDERED FIELDS

**1. THE FIELD OF RATIONAL NUMBERS.** In the integral domain **Z**, the only elements having multiplicative inverses are ±1. (See Exercise 2, 4.3.) Using **Z**, we now construct a ring **Q** with the property that every non-zero element of **Q** has a multiplicative inverse. The elements of **Q** are the rational numbers.

Our definition of rational numbers will be based on familiar properties of fractions. We observe that two fractions (e.g., 2/4 and −3/−6) are equal when their "cross products" are equal: 2(−6) = (−3)4. We shall define a rational number as an equivalence class of ordered pairs of integers, where the equivalence relation is given by equality of cross products.

THEOREM 1.1 *If* $A = \mathbf{Z} \times \mathbf{Z} - \{0\}$, *then there is an equivalence relation* $Q$ *in* $A$ *such that* $(a, b)\, Q\, (c, d)$ *holds whenever* $ad = cb$.

PROOF: The set

$$Q = \{((a, b), (c, d)) \mid ad = cb \text{ in } \mathbf{Z}\}$$

is a subset of $A \times A$. By Definition 0.6, $Q$ is a binary relation in $A$. Since $ab = ab$, $((a, b), (a, b)) \in Q$ for all $(a, b) \in A$. Hence the relation, $Q$, is reflexive. If $((a, b), (c, d)) \in Q$, then $((c, d), (a, b)) \in Q$. Hence $Q$ is symmetric. Finally, if $((a, b), (c, d))$ and $((c, d), (e, f))$ are in $Q$, then $ad = cb$ and $cf = ed$. By the properties of multiplication in **Z**, $adf = cbf = edb$ and, since $d \neq 0$, $af = eb$. Hence, $((a, b), (e, f)) \in Q$, and the relation, $Q$, is transitive. By Definition 0, 3.3, $Q$ is an equivalence relation in $A$.

DEFINITION 1.1 We write "$(a, b) \sim (c, d)$" for "$((a, b), (c, d)) \in Q$" and read "$\sim$" as "is equivalent to." For each $(a, b) \in A$, the equivalence class $C_{(a,b)}$ is the set of all $(c, d) \in A$ such that $(c, d) \sim (a, b)$. A *rational number* is an equivalence class $C_{(a,b)}$. We write "**Q**" for "the set of all rational numbers" and "$x$," "$y$," "$z$," . . . for elements of **Q**.

If $A$ and $Q$ are the sets defined in Theorem 1.1, then the factor set, $A/Q$, is the set of all rational numbers.

Our definitions of addition and multiplication in $\mathbf{Q}$ mirror the properties of the familiar addition and multiplication of fractions.

THEOREM 1.2 *If* $(a, b) \sim (a', b')$ *and* $(c, d) \sim (c', d')$, *then*

(1) $(ad + cb, bd) \sim (a'd' + c'b', b'd')$,

*and*

(2) $(ac, bd) \sim (a'c', b'd')$.

PROOF: By hypothesis, $ab' = a'b$ and $cd' = c'd$. Hence, by the properties of addition and multiplication in $\mathbf{Z}$,

$$\begin{aligned}(ad + cb)(b'd') &= (ab')(dd') + (cd')(bb') \\ &= (a'b)(dd') + (c'd)(bb') \\ &= (a'd' + c'b')(bd).\end{aligned}$$

Since $bd \neq 0$ and $b'd' \neq 0$, $(ad + cb, bd) \sim (a'd' + c'b', b'd')$ by Definition 1.1. Also, $(ac)(b'd') = (ab')(cd') = (a'b)(c'd) = (a'c')(bd)$, and, since $bd \neq 0$ and $b'd' \neq 0$, $(ac, bd) \sim (a'c', b'd')$.

THEOREM 1.3 *There are binary operations F and G on* $\mathbf{Q}$ *such that, if* $(a, b) \in x$ *and* $(c, d) \in y$, *then*

(1) $F(x, y) = C_{(ad+cb, bd)}$,

(2) $G(x, y) = C_{(ac, bd)}$.

PROOF: The sets

$$F = \{((x, y), C_{(ad+cb, bd)}) \mid (a, b) \in x, (c, d) \in y; x, y \in \mathbf{Q}\}$$

and

$$G = \{((x, y), C_{(ac, bd)}) \mid (a, b) \in x, (c, d) \in y; x, y \in \mathbf{Q}\}$$

are subsets of $(\mathbf{Q} \times \mathbf{Q}) \times \mathbf{Q}$. Since $(a, b)$ and $(c, d)$ are in $A$, $(ad + cb, bd)$ and $(ac, bd)$ are in $A$.

For each $(x, y) \in \mathbf{Q} \times \mathbf{Q}$, there are $(a, b) \in x$, $(c, d) \in y$, and $z = C_{(ad+cb, bd)}$ such that $((x, y), z) \in F$. If $(a', b') \in z$, $(c', d') \in y$, and $z' = C_{(a'd'+c'b', b'd')}$, then, from $(a', b') \sim (a, b)$, $(c', d') \sim (c, d)$, and Theorem 1.2, it follows that $(ad + cb, bd) \sim (a'd' + c'b', b'd')$. Hence, by Definition 1.1, $z' = z$. Therefore, $F$ is a function from $\mathbf{Q} \times \mathbf{Q}$ to $\mathbf{Q}$ by Definition 0, 4.1, and, by Definition 1, 3.1, $F$ is a binary operation on $\mathbf{Q}$.

For each $(x, y) \in \mathbf{Q} \times \mathbf{Q}$, there are $(a, b) \in x$, $(c, d) \in y$, and $z = C_{(ac, bd)}$ such that $((x, y), z) \in G$. If $(a', b') \in x$, $(c', d') \in y$, and $z' = C_{(a'c', b'd.)}$; then $(a', b') \sim (a, b)$, $(c', d') \sim (c, d)$, and, by Theorem 1.2, $(ac, bd) \sim (a'c', b'd')$.

Hence, $z = z'$ by Definition 1.1. Therefore, $G$ is a function from $\mathbf{Q} \times \mathbf{Q}$ to $\mathbf{Q}$. By Definition 1, 3.1, $G$ is a binary operation on $\mathbf{Q}$.

DEFINITION 1.2 We refer to the binary operations $F$ and $G$ of Theorem 1.3, respectively, as *addition* and *multiplication in* $\mathbf{Q}$ and write

$$x +_{\mathbf{Q}} y = F(x, y), \qquad x \cdot_{\mathbf{Q}} y = G(x, y)$$

for all $x, y \in \mathbf{Q}$.

We omit the subscript "$\mathbf{Q}$" and even the "$\cdot$" if the context makes it clear that the operations are in $\mathbf{Q}$.

THEOREM 1.4 $\langle \mathbf{Q}, +_{\mathbf{Q}}, \cdot_{\mathbf{Q}} \rangle$ *is a commutative ring with identity.*

PROOF:

(1) Addition in $\mathbf{Q}$ is commutative.

If $x, y \in \mathbf{Q}$, then $x = C_{(a,b)}, y = C_{(c,d)}$ for some $a, b, c, d \in \mathbf{Z}$, $b \neq 0, d \neq 0$. By Definition 1.2, Theorem 1.3, and the properties of addition and multiplication in $\mathbf{Z}$,

$$\begin{aligned} x +_{\mathbf{Q}} y = F(x, y) &= C_{(ad+cb,bd)} \\ &= C_{(cb+ad,db)} = F(y, x) = y +_{\mathbf{Q}} x. \end{aligned}$$

(2) Addition in $\mathbf{Q}$ is associative.

If $x, y, z \in \mathbf{Q}$, then $x = C_{(a,b)}, y = C_{(c,d)}, z = C_{(e,f)}$, for some $(a, b), (c, d)$, and $(e, f)$ in $\mathbf{Z} \times \mathbf{Z} - \{0\}$. Now, by Definition 1.2, Theorem 1.3, and the properties of addition and multiplication in $\mathbf{Z}$,

$$(1) \quad \begin{gathered} (ad + cb, bd) \in x +_{\mathbf{Q}} y, \\ ((ad + cb)f + e(bd), (bd)f) \in (x +_{\mathbf{Q}} y) +_{\mathbf{Q}} z, \end{gathered}$$

$$(2) \quad \begin{gathered} (cf + ed, df) \in y +_{\mathbf{Q}} z, \\ (a(df) + (cf + ed)b, b(df)) \in x +_{\mathbf{Q}} (y +_{\mathbf{Q}} z). \end{gathered}$$

Since, in $\mathbf{Z}$,

$$(ad + cb)f + e(bd) = a(df) + (cf + ed)b \text{ and } (bd)f = b(fd),$$

it follows from (1) and (2) that the equivalence classes $(x +_{\mathbf{Q}} y) +_{\mathbf{Q}} z$ and $x +_{\mathbf{Q}} (y +_{\mathbf{Q}} z)$ have an element in common. Hence,

$$x +_{\mathbf{Q}} (y +_{\mathbf{Q}} z) = (x +_{\mathbf{Q}} y) +_{\mathbf{Q}} z.$$

(3) $\mathbf{Q}$ contains a unique identity for addition. If $x = C_{(a,b)}$ is an element of $\mathbf{Q}$, then

$$x + C_{(0,1)} = C_{(a \cdot 1 + 0 \cdot b, b \cdot 1)} = C_{(a,b)} = x.$$

Hence, $C_{(0,1)}$ is an identity for addition. By Theorem 2, 3.1, there is only one identity.

(4) Every element of **Q** has a unique inverse with respect to addition. If $x = C_{(a,b)} \in \mathbf{Q}$ and $x' = C_{(-a,b)}$, then

$$x +_{\mathrm{Q}} x' = C_{(ab+(-a)b,b^2)} = C_{(0,b^2)} = C_{(0,1)},$$

since $(0, b^2) \sim (0, 1)$. Hence $x'$ is an inverse for $x$. By Theorem 2, 3.2, $x'$ is the only inverse. We leave it to the reader to verify that multiplication is associative and commutative, and note that the element $C_{(1,1)}$ serves as an identity for multiplication in **Q**.

It remains only to be proved that multiplication in **Q** distributes over addition.

If $x = C_{(a,b)}, y = C_{(c,d)}$, and $z = C_{(e,f)}$, then

$$\begin{aligned} x \cdot_{\mathrm{Q}} y +_{\mathrm{Q}} x \cdot_{\mathrm{Q}} z &= C_{(ac,bd)} +_{\mathrm{Q}} C_{(ae,bf)} \\ &= C_{(acbf+aebd,b^2df)} \\ &= C_{(acf+aed,bdf)} \\ &= C_{(a,b)} \cdot_{\mathrm{Q}} (C_{(c,d)} +_{\mathrm{Q}} C_{(e,f)}) = x \cdot_{\mathrm{Q}} (y +_{\mathrm{Q}} z) \end{aligned}$$

since $b \neq 0$ in **Z**.

*Notation:* We write $0_{\mathrm{Q}}$ for $C_{(0,1)}$ and $1_{\mathrm{Q}}$ for $C_{(1,1)}$.

THEOREM 1.5 *Every element x in* **Q** *other than* 0 *has a unique inverse with respect to multiplication.*

PROOF: If $x = C_{(a,b)} \in \mathbf{Q}$ and $x \neq 0_{\mathrm{Q}} = C_{(0,1)}$, then $a \neq 0$, since $(0, b) \sim (0, 1)$ for all $b \neq 0$ in **Z**. If $a \neq 0$, then $x' = C_{(b,a)} \in \mathbf{Q}$ and

$$x \cdot_{\mathrm{Q}} x' = G(x, x') = C_{(ab,ba)} = C_{(1,1)} = 1_{\mathrm{Q}}.$$

Hence, $x'$ is an inverse for $x$ with respect to multiplication in **Q**. By Theorem 2.6, $x'$ is the only inverse of $x$.

DEFINITION 1.3 Let $\langle A, +, \cdot \rangle$ be a commutative ring with identity $1 \neq 0$. Then $\langle A, +, \cdot \rangle$ is a *field* if every non-zero element of $A$ has a multiplicative inverse in $A$.

*Remark:* Every field is an integral domain (see Exercise 1.2). A ring $A$ is a field if and only if the non-zero elements of $A$ form an abelian group (see Exercise 1.3).

THEOREM 1.6 $\langle \mathbf{Q}, +_{\mathrm{Q}}, \cdot_{\mathrm{Q}} \rangle$ *is a field.*

PROOF: This follows immediately from Theorems 1.4 and 1.5.

*Notation:* If $\langle A, +, \cdot\rangle$ is a field, we write "$0_A$" and "$1_A$" (or simply "0," "1") for the additive and multiplicative identities, respectively. If $x \neq 0$, we write "$1/x$" for the multiplicative inverse of $x$ in $A$, and, generally, "$y/x$" for the element $z \in A$ such that $xz = y$.

• *Exercise 1.1* Complete the proof of Theorem 1.2.

• *Exercise 1.2* Prove that every field is an integral domain. Note that **Z** is an integral domain but not a field.

• *Exercise 1.3* Prove: a ring $A$ is a field if and only if the non-zero elements of $A$ form an abelian group.

*Exercise 1.4* Let $A$ be an integral domain. Prove that, by a construction similar to that of **Q** from **Z**, a field can be constructed from $A$. (This field is called the *quotient field* of $A$.)

*Exercise 1.5* The ring $\langle A_k, \oplus, \odot\rangle$ defined in Exercise 2, 8.10 is a field if and only if $k$ is prime.

*Exercise 1.6* The properties of commutative groups asserted in Exercise 2, 3.6 in additive notation, applied to the group $\langle A', \cdot\rangle$, where $A' = A - \{0\}$, read as follows:

For all $x, y, z \in A'$,

(1) $1/(1/x) = x$,
(2) $x \cdot (1/y) = x/y$,
(3) $1/xy = (1/x)(1/y) = (1/x)/y$,
(4) $(x/y)(y/z) = x/z$,
(5) $1/(x/y) = y/x$.

Also, for all $x, z \in A, y, t \in A'$,

$$x/y + z/t = (xt + zy)/yt.$$

**2. ORDER.** As in the case of **Z**, we begin by introducing a set of positive elements.

*Notation:* We let $\mathbf{Q}^+ = \{x \in \mathbf{Q} \mid ab >_{\mathbf{Z}} 0_{\mathbf{Z}} \text{ for some } (a, b) \in x\}$.

THEOREM 2.1 *If $ab >_{\mathbf{Z}} 0_{\mathbf{Z}}$ and $(a, b) \sim (c, d)$, then $cd >_{\mathbf{Z}} 0_{\mathbf{Z}}$.*

PROOF: From $ad = cb$, we have $(cb)(ad) = (cb)(cb)$.

Hence, $(ab)(cd) = (cb)(cb)$. By Exercise 2, 7.3, $(cb)(cb) >_{\mathbf{Z}} 0_{\mathbf{Z}}$ in **Z**. But then $(ab)(cd) >_{\mathbf{Z}} 0_{\mathbf{Z}}$ in **Z** and, since $ab >_{\mathbf{Z}} 0_{\mathbf{Z}}$ in **Z**, we have $cd >_{\mathbf{Z}} 0_{\mathbf{Z}}$ in **Z** (Theorem 2, 6.5).

COROLLARY $\mathbf{Q}^+ = \{x \in \mathbf{Q} \,|\, ab > 0 \text{ for all } (a, b) \in x\}$.

THEOREM 2.2 *$\mathbf{Q}^+$ is a set of positive elements for $\mathbf{Q}$.*

PROOF: We show that $\mathbf{Q}^+$ satisfies (1), (2), and (3) of Definition 2, 7.3. If $x, y \in \mathbf{Q}^+$, then $z = C_{(a,b)}, y = C_{(c,d)}$, where $a, b, c, d \in \mathbf{Z}$, and

$$(1) \qquad ab >_{\mathbf{Z}} 0_{\mathbf{Z}}, \qquad cd >_{\mathbf{Z}} 0_{\mathbf{Z}}.$$

Hence,

$$x + y = C_{(a,b)} + C_{(c,d)} = C_{(ad+cb,bd)}.$$

By Theorem 2, 6.3,

$$(ad + cb)bd = abdd + cdbb >_{\mathbf{Z}} 0_{\mathbf{Z}}.$$

Hence, $x + y \in \mathbf{Q}^+$, and (1) of Definition 2.10 is fulfilled.

Also, $x \cdot y = C_{(a,b)} \cdot C_{(c,d)} = C_{ac,bd)}$, and $(ac)(bd) >_{\mathbf{Z}} 0_{\mathbf{Z}}$, so that $x \cdot y \in \mathbf{Q}^+$, and (2) of Definition 2, 7.3 is fulfilled.

Finally, if $x = C_{(a,b)}$, then, by the trichotomy property of order in $\mathbf{Z}$, exactly one of

$$ab >_{\mathbf{Z}} 0_{\mathbf{Z}} \qquad ab = 0_{\mathbf{Z}} \qquad -ab >_{\mathbf{Z}} 0_{\mathbf{Z}}$$

must hold. But, by Theorem 2.1,

$$ab >_{\mathbf{Z}} 0_{\mathbf{Z}} \text{ if and only if } x \in \mathbf{Q}^+,$$

$$(-a) \cdot b = -ab >_{\mathbf{Z}} 0_{\mathbf{Z}} \text{ if and only if } C_{(-a,b)} = -x \in \mathbf{Q}^+,$$

and, since $b \neq 0_{\mathbf{Z}}$ and $\mathbf{Z}$ is an integral domain,

$$ab = 0_{\mathbf{Z}} \text{ if and only if } a = 0_{\mathbf{Z}}, \text{ and } x = C_{(0,b)} = 0_{\mathbf{Q}}.$$

Thus, $\mathbf{Q}^+$ satisfies (3) of Definition 2, 7.3.

By Theorem 2.3 (1), we have

THEOREM 2.3 *The set $T = \{(x, y) \,|\, y - x \in \mathbf{Q}^+\}$ is an order relation in $\mathbf{Q}$.*

*Notation:* We write $x <_{\mathbf{Q}} y$ ($y >_{\mathbf{Q}} x$) if $(x, y) \in T$. We usually omit the subscript $\mathbf{Q}$.

By Theorem 2, 7.2 (2) we have

THEOREM 2.4 *$\langle \mathbf{Q}, +_{\mathbf{Q}}, \cdot_{\mathbf{Q}} <_{\mathbf{Q}} \rangle$ is an ordered integral domain.*

DEFINITION 2.1 If $\langle A, +, \cdot, < \rangle$ is an ordered integral domain such that $\langle A, +, \cdot \rangle$ is a field, then $\langle A, +, \cdot, < \rangle$ is called an *ordered field.*

**3. EMBEDDING.** We now show that the ordered field $\langle \mathbf{Q}, +_{\mathbf{Q}}, \cdot_{\mathbf{Q}} <_{\mathbf{Q}} \rangle$ is an extension of the ordered integral domain $\langle \mathbf{Z}, +_{\mathbf{Z}}, \cdot_{\mathbf{Z}}, <_{\mathbf{Z}} \rangle$.

THEOREM 3.1 *If $E$: $\mathbf{Z} \to \mathbf{Q}$ is defined by: $E(a) = C_{(a,1)}$ for each $a \in \mathbf{Z}$, then $E$ is an isomorphic embedding with respect to addition, multiplication and order.*

PROOF: $E$ is an injection. For, if $E(a) = E(b)$, then $C_{(a,1)} = C_{(b,1)}$; hence, $a \cdot 1 = b \cdot 1$, and $a = b$.

For $a, b \in \mathbf{Z}$, $E(a + b) = C_{(a+b,1)} = C_{(a,1)} + C_{(b,1)} = E(a) + E(b)$. Thus, $E$ preserves multiplication.

For $a, b \in \mathbf{Z}$, $E(ab) = C_{(ab,1)} = C_{(a,1)}C_{(b,1)} = E(a) \cdot E(b)$. Thus, $E$ preserves multiplication.

For $a, b \in \mathbf{Z}$, $a <'_{\mathbf{Z}} b$ if and only if $b - a \in \mathbf{Z}^+$, i.e., if and only if $C_{(b-a,1)} = C_{(b,1)} - C_{(a,1)} = E(b) - E(a) \in \mathbf{Q}^+$, and $E(a) <_{\mathbf{Q}} E(b)$.

*Notations:* In the following we refer to the embedding of $\mathbf{N}$ in $\mathbf{Z}$ (Theorem 2, 8.1) as $E_{\mathbf{N}}^{\mathbf{Z}}$ and to the embedding of $\mathbf{Z}$ in $\mathbf{Q}$ as $E_{\mathbf{Z}}^{\mathbf{Q}}$. In view of the isomorphic embeddings $E_{\mathbf{N}}^{\mathbf{Z}}$ and $E_{\mathbf{Z}}^{\mathbf{Q}}$, we shall from now on use the same notation for elements of $\mathbf{Z}$ and their images in $\mathbf{Q}$. Since we have previously used "$n$" to denote $E_{\mathbf{N}}^{\mathbf{Z}}(n) = C_{(S(n),1)}$ in $\mathbf{Z}$, we now use "$n$" to denote $E_{\mathbf{Z}}^{\mathbf{Q}}(E_{\mathbf{N}}^{\mathbf{Z}}(n)) = C_{(n,1)}$ in $\mathbf{Q}$. We note that $E_{\mathbf{Z}}^{\mathbf{Q}}(E_{\mathbf{N}}^{\mathbf{Z}}(1) = 1_{\mathbf{Q}}$, and $E_{\mathbf{Z}}^{\mathbf{Q}}(0) = 0_{\mathbf{Q}}$, so that we are justified in writing 1 and 0 for $1_{\mathbf{Q}}$ and $0_{\mathbf{Q}}$, respectively. Finally, since for $h \in \mathbf{Z}$, $k \neq 0$ in $\mathbf{Z}$,

$$C_{(h,k)} = C_{(h,1)}/C_{(k,1)} = E_{\mathbf{Z}}^{\mathbf{Q}}(h)/E_{\mathbf{Z}}^{\mathbf{Q}}(k),$$

we write $h/k$ for the rational number $C_{(h,k)}$.

We also use "$\mathbf{N}$" and "$\mathbf{Z}$" to designate, respectively, the images of $\mathbf{N}$ and $\mathbf{Z}$ in $\mathbf{Q}$.

• *Exercise 3.1* If $F$ is a function from an ordered field $A$ to an ordered field $B$ which preserves addition and maps positive elements to positive elements, then $F$ preserves order.

• *Exercise 3.2* If $x$, $y$ are rational numbers, then $x = h/n$, $y = k/n$ for some $h, k \in \mathbf{Z}$, $n \in \mathbf{Z}^+$. If $x$, $y$ are positive rational numbers, then $x = p/n$, $y = q/n$ for some $p, q, n \in \mathbf{Z}^+$.

*Exercise 3.3* Show that $\mathbf{Q}$ is countable, i.e., prove that there is a bijection from $\mathbf{N}$ to $\mathbf{Q}$. Is there a bijection from $\mathbf{N}$ to $\mathbf{Q}$ which preserves addition or multiplication?

**4. PROPERTIES OF ORDERED FIELDS.** The ordered field of rational numbers plays a fundamental role among all ordered fields.

THEOREM 4.1 *The ordered field* **Q** *of rational numbers can be ismorphically embedded in any ordered field.*

PROOF: Let $A$ be an ordered field. First define $F\colon \mathbf{Z} \to A$ as follows:

$$F(1) = 1_A$$

$$F(n+1) = F(n) + 1_A \qquad \text{for each } n \in \mathbf{Z}^+$$

$$F(0) = 0_A$$

$$F(-n) = -F(n) \qquad \text{for each } n \in \mathbf{Z}^+.$$

Then $F$ is a monomorphism preserving addition, multiplication and order. For each $k \in \mathbf{Z}$, denote $F(k)$ by $k1_A$. Now extend $F$ to $G\colon \mathbf{Q} \to A$ by letting

$$G\left(\frac{h}{k}\right) = \frac{h1_A}{k1_A}$$

for each $h/k \in \mathbf{Q}$. Then $G$ is a monomorphism of **Q** into $A$ preserving addition, multiplication and order. (We leave the details to the reader—see Exercise 4.1.)

DEFINITION 4.1 We call the elements $h1_A/k1_A$ of an ordered field $A$ *the rational elements of* $A$ and denote by "$\mathbf{Q}_A$" the set of all rational elements of $A$.

For any ordered field $\langle A, +, \cdot, < \rangle$, there is a function from $A$ to $A$ which may be used to define *distance* in $A$.

DEFINITION 4.2 If $\langle A, +, \cdot, < \rangle$ is an ordered field, and $F\colon A \to A$ is defined by

$$F(x) = \max\{x, -x\},$$

then for $x \in A$, $F(x)$ is called the *absolute value* of $x$, and is denoted by $|x|$.

(Note: By the trichotomy of the order relation in $A$, there is such a function.)

In the following, $\langle A, +, \cdot, < \rangle$ is an ordered field.

THEOREM 4.2 *Let $A$ be an ordered field, $x, y \in A$. Then:*

(1) a. $|x| = |-x| \geqslant 0$,

b. $x \leqslant |x|$ and $-x \leqslant |x|$,

c. $|x| = 0$ if and only if $x = 0$;

(2) $|x+y| \leqslant |x| + |y|$,
(3) $|xy| = |x| \cdot |y|$,
(4) $|x| - |y| \leqslant ||x| - |y|| \leqslant |x-y|$.

PROOF: We leave to the reader the proofs of (1) and (3). (See Exercise 4.2)

(2) By (1)b. and Exercise 2, 7.5, we have $x+y \leqslant |x| + |y|$ and $-(x+y) \leqslant |x| + |y|$. By Definition 4.2, $|x+y|$ is one of $x+y, -(x+y)$. Hence $|x+y| \leqslant |x| + |y|$.

(4) Let $z = x - y$. Then $x = z + y$, hence, by (2), we have $|x| = |z+y| \leqslant |z| + |y|$, and $|x| - |y| \leqslant |z| = |x-y|$. Similarly, since $-z = y - x$, we have $|y| - |x| \leqslant |-z| = |z| = |x-y|$. But then, by Definition 4.2, we conclude that

$$|x| - |y| \leqslant ||x| - |y|| \leqslant |x-y|.$$

*Remark:* If $A$ is an ordered field and, for $x, y \in A$, the distance from $x$ to $y$ is defined by

$$d(x, y) = |x-y|,$$

then

$$d(x, y) = 0 \quad \text{if and only if } x = y,$$

$$d(x, y) = d(y, x) \quad \text{for all } x, y \in A,$$

$$d(x, z) \leqslant d(x, y) + d(y, z) \quad \text{for all } x, y, z \in A.$$

These are properties of the familiar notion of distance. (See Exercise 4.4 and Chapter 4, Section 5.)

In the ordered integral domain, **Z**, there are *consecutive elements:* if $a$ is any integer, then there is no integer between $a$ and $a + 1$. In an ordered field, on the other hand, there is always another element of the field between any two given elements.

DEFINITION 4.3 An order relation $<$ in a set $A$ is called *dense* if, for any $a$, $b$ such that $a < b$ in $A$, there is some $c \in A$ such that $a < c < b$.

THEOREM 4.3 *If $\langle A, +, \cdot, < \rangle$ is an ordered field, then the order in $A$ is dense.*

PROOF: If $a < b$ in $A$, then

$$2a = a + a < a + b < b + b = 2b.$$

Since $2 = 1_A + 1_A > 0_A$ in $A$, it follows that $1/2 > 0$ in $A$, and $a < (a+b)/2 < b$.

COROLLARY *Between any two rational numbers, there is another rational number.*

Another important property of the order in **Q** is that every positive element has arbitrarily large integral multiples. This is not true in all ordered fields, as we shall see.

DEFINITION 4.4 An ordered field $\langle A, +, \cdot\rangle, <$ is called Archimedean if, for any $a, b \in A$ such that $0 < a < b$ in $A$, there is some $n \in \mathbf{N}$ such that $na \geqq b$.

THEOREM 4.4 *$\langle \mathbf{Q}, +, \cdot, < \rangle$ is Archimedean.*

PROOF: If $0 < x < y$, then (Exercise 3.2) there are positive integers $p$, $q$, $r$ such that $x = p/r$, $y = q/r$. Then

$$0 < \frac{p}{r} < \frac{q}{r}$$

and

$$\frac{p}{r} \cdot rq = pq \geqq q \geqq \frac{q}{r}.$$

Thus, for $n = rq$, $nx \geqq y$.

THEOREM 4.5 *An ordered field, $A$, is Archimedean if and only if, for each $e > 0$ in $A$, there is some $n \in \mathbf{N}$ such that $\frac{1}{n} < e$.*

PROOF: Suppose $A$ is Archimedean. If $e \geqslant 1$, then $\frac{1}{n} < 1 \leqslant e$ for each $n \geqslant 2$. If $0 < e < 1$ then, since $A$ is Archimedean, there is some $n \in \mathbf{N}$ such that $ne > 1$, hence $\frac{1}{n} < e$.

Conversely, suppose $A$ is an ordered field such that, for each $e > 0$ in $A$, there is some $n \in \mathbf{N}$ such that $\frac{1}{n} < e$. Let $a, b \in A$, $0 < a < b$, and let $e = \frac{a}{b}$. By hypothesis, there is some $n \in \mathbf{N}$ such that $\frac{1}{n} < \frac{a}{b} = e$. But then $na > b$, and so $A$ is Archimedean.

• *Exercise 4.1* Complete the proof of Theorem 4.1.

• *Exercise 4.2* Complete the proof of Theorem 4.2.

*Exercise 4.3* Between any two elements of an ordered field $A$, there are infinitely many elements of $A$.

• *Exercise 4.4* Let $A$ be an ordered field. For $x, y \in A$, let $d(x, y) = |x - y|$. Prove:

(1) $d(x, y) = 0$ if and only if $x = y$.
(2) $d(x, y) = d(y, x)$ for all $x, y \in A$.
(3) $d(x, z) \leqslant d(x, y) + d(y, z)$ for all $x, y, z \in A$.

*Exercise 4.5*

(a) Let $A$ be the set of all polynomials $a_0 + a_1x + \cdots + a_nx^n$, where $n \geqslant 0$ in $\mathbf{Z}$ and $a_i \in \mathbf{Q}$ for $i = 0, \ldots, n$. On $A$, define $+, \cdot$ as the usual addition and multiplication of polynomials, i.e.,

$$\sum_{i=0}^{n} a_ix^i + \sum_{i=0}^{n} b_ix^i = \sum_{i=0}^{n} (a_i + b_i)x^i,$$

and

$$\sum_{i=0}^{n} a_ix^i \cdot \sum_{j=0}^{m} b_jx^j = \sum_{k=0}^{m+n}\left(\sum_{i+j=k} a_ib_j\right)x^k.$$

Verify that $\langle A, +, \cdot\rangle$ is an integral domain.

(b) Let $A^+$ be the set of all non-zero polynomials in $A$ whose first non-zero coefficient is positive. Prove that $A^+$ is a set of positive elements for $A$, in the sense of Definition 2, 7.3.

(c) Define $<$ in $A$ by: $f < g$ if $g - f \in A^+$. Conclude that $\langle A, +, \cdot, <\rangle$ is an ordered integral domain. Prove that $<$ is non-Archimedean. (Hint: consider the polynomials $f = 1$ and $g = x$.)

(d) Let $K = \{\frac{f}{g} | f, g \in A, g \neq 0\}$. Define $+, \cdot$ on $K$ by:

$$\frac{f_1}{g_1} + \frac{f_2}{g_2} = \frac{f_1g_2 + f_2g_1}{g_1g_2}$$

$$\frac{f_1}{g_1} \cdot \frac{f_2}{g_2} = \frac{f_1f_2}{g_1g_2}$$

and define $K^+ = \{\frac{f}{g} | f, g \in A \text{ and } fg \in A^+\}$. Prove that $K^+$ is a set of positive elements for $K$. Let $<$ be the order relation associated with $K^+$. Prove that $\langle K, +, \cdot, <\rangle$ is a non-Archimedean ordered field.

(e) Let $f = a_0 + a_1x + \cdots + a_nx^n$ $(a_i \in A, i = 0, \ldots, n)$

$g = b_0 + b_1x + \cdots + b_mx^m$ $(b_j \in A, j = 0, \ldots, m, b_m \neq 0)$.

If the quotient $\frac{f}{g}$ is "divided out," then a "Laurent series" of form $c_{-k}x^{-k} + c_{-k-1}x^{-k-1} + \cdots + c_0x^0 + c_1x + c_2x^2 + \cdots$ is obtained. Prove that $\frac{f}{g}$ is positive in $K$ if and only if the first non-zero coefficient of $\sum_{-k}^{\infty} c_ix^i$ is positive in $\mathbf{Q}$.

5. **GAPS IN Q; SEQUENCES IN ORDERED FIELDS.** Despite its density, there are *gaps* in the order of **Q**, in the sense of the following theorem.

THEOREM 5.1

(1) *For each* $x \in \mathbf{Q}$, $x^2 \neq 2$ (*i.e.,* $\mathbf{Q}$ *contains no square root of* 2).

(2) *Let* $X = \{x \in \mathbf{Q} \mid x < 0 \text{ or } x^2 < 2\}$,

$$Y = \{y \in \mathbf{Q} \mid y > 0 \text{ and } y^2 > 2\}.$$

*Then* (a) $X \neq \phi$ and $Y \neq \phi$;

(b) $X \cap Y = \phi$ and $X \cup Y = \mathbf{Q}$;

(c) *if* $x \in X$ and $y \in Y$, *then* $x < y$;

(d) *X has no maximum and Y has no minimum.*

PROOF: (1) Suppose there is a rational number whose square is 2. Since $y^2 = (-y)^2$ for each $y \in \mathbf{Q}$, there is, in particular, a *positive* rational number whose square is 2. Hence, by Exercise 3.2, there are positive integers $a$ and $b$ such that $(\frac{a}{b})^2 = 2$, whence $a^2 = 2b^2$. By the well-ordering of $\mathbf{Z}^+$, there is a least positive integer $a$ such that $a^2 = 2b^2$ for some $b \in \mathbf{Z}^+$. Since $a$ and $b$ are positive, the relation $a^2 = 2b^2$ implies $b^2 < a^2$, hence $b < a$. Also, from $a^2 = 2b^2$, we infer that $a^2$ is even, hence $a$ is even (see Exercise 2, 8.8). But then $a = 2c$ for some $c \in \mathbf{Z}^+$, and $a^2 = 4c^2 = 2b^2$ yields $b^2 = 2c^2$. Since $b < a$, this contradicts the minimal property of $a$. Thus, there is no $x \in \mathbf{Q}$ such that $x^2 = 2$.

(2) We leave to the reader the verification of (a), (b) and (c) (see Exercise 5.1).

(d) Suppose $X$ has a maximum, $x_0$. Since $1 \in X$, $x_0 \geqslant 1 > 0$, hence

$$\frac{2 - x_0^2}{2x_0 + 1} > 0.$$

By the density of order in $\mathbf{Q}$, there is a rational number $e$ such that

$$0 < e < \min\left\{1, \frac{2 - x_0^2}{2x_0 + 1}\right\}.$$

Then $(x_0 + e)^2 = x_0^2 + 2x_0e + e^2 \leqslant x_0^2 + e(2x_0 + 1) < x_0^2 + 2 - x_0^2 = 2$. This contradicts our hypothesis that $x_0 = \max X$.

Next, suppose $Y$ has a minimum, $y_0$. Then $y_0 > 0$ and

$$\frac{y_0^2 - 2}{2y_0} > 0.$$

By the density of order in $\mathbf{Q}$, there is a rational number $e$ such that

$$0 < e < \frac{y_0^2 - 2}{2y_0}.$$

Then $(y_0 - e)^2 = y_0{}^2 - 2y_0e + e^2 \geqslant y_0{}^2 - 2y_0e > 2$. This contradicts our hypothesis that $y_0 = \min Y$.

DEFINITION 5.1 If $\langle A, +, \cdot, < \rangle$ is an ordered field, then the ordered pair $(X, Y)$ is a *cut* in $A$ if $X$ and $Y$ are non-empty subsets of $A$ such that

(1) $X \cap Y = \phi$,

(2) $X \cup Y = A$,

(3) if $x \in X, y \in Y$, then $x < y$.

The sets $X$ and $Y$ are called, respectively, the *lower class* and the *upper class* of the cut.

A cut is a *gap* if its lower class has no greatest element, and its upper class has no smallest element, in $A$.

We note that the ordered pair $(X, Y)$ of Theorem 5.1 is a cut and, in fact, a gap, in **Q** due to the absence in **Q** of a square root of 2. In Chapter 4, we shall embed the ordered field **Q** in an ordered field **R** in which there are no gaps. This will be the field of real numbers. In the construction of **R**, we shall employ sequences of rational numbers.

We recall (Definition 1, 2.1) that a sequence in a set $A$ is a function $F: \mathbf{N} \to A$. If $F(n) = a_n$, for each $n \in \mathbf{N}$, we use the symbol $(a_n)$ to designate the sequence $F$.

The following theorem shows that every gap in **Q** is "quite narrow" and can be "approximated" by sequences of rational numbers. In extending **Q** to the field of real numbers, we fill each gap by an equivalence class whose elements are the sequences which approximate the gap.

THEOREM 5.2 *If $(X, Y)$ is a gap in **Q**, then there are sequences $(x_n)$, $(y_n)$ in **Q** such that for each $n \in N$,*

$$x_n \in X, y_n \in Y$$

$$y_n - x_n = \frac{1}{n}$$

*and*

$$|x_m - x_n| < \frac{1}{n}, \; |y_m - y_n| < \frac{1}{n} \text{ in } \mathbf{Q}$$

*for all $m \geqq n$ in **N**.*

PROOF: Since $(X, Y)$ is a cut, $X$ and $Y$ are not empty. If $x \in X$ and $y \in Y$, then $y - x > 0$ in **Q**. For each $n \in \mathbf{N}$, $1/n > 0$ in **Q** and, since the order in **Q** is Archimedean, there is some $k_n \in \mathbf{N}$ such that $k_n/n \geqq y - x$ in **Q**. Since $y \in Y$ and $x + k_n/n \geqq y$, $x + k_n/n$ is an element of $Y$. Hence, for each $n \in \mathbf{N}$, the set

$$M_n = \left\{ m \in \mathbf{N} \,\middle|\, x + \frac{m}{n} \in Y \right\}$$

is a non-empty subset of **N** and contains a first element $m_n$ (Theorem 1, 4.5). Now, for each $n \in \mathbf{N}$,

$$x_n = x + \frac{m_n - 1}{n} \in X, \quad y_n = x + \frac{m_n}{n} \in Y,$$

and

$$y_n - x_n = \frac{1}{n}.$$

Since $(X, Y)$ is a cut in **Q**, $x_n < y_m$ in **Q** for all $m, n \in \mathbf{N}$. Hence

$$x_n < y_m = x_m + \frac{1}{m} \text{ and } x_m < y_n = x_n + \frac{1}{n} \text{ for all } m, n \in \mathbf{N}.$$

Therefore, for each $n \in \mathbf{N}$ and all $m \geqq n$ in **N**,

$$|x_m - x_n| = \max\{x_m - x_n, x_n - x_m\} < \max\left\{\frac{1}{n}, \frac{1}{m}\right\} = \frac{1}{n}$$

and

$$|y_m - y_n| = \max\{y_m - y_n, y_n - y_m\} < \max\left\{\frac{1}{m}, \frac{1}{n}\right\} = \frac{1}{n}$$

in **Q**.

COROLLARY: *The sequence* $(x_n)$ *of Theorem 5.2 has the property: for every* $e > 0$ *in* **Q**, *there is some* $n_e \in \mathbf{N}$ *such that*

$$|x_m - x_n| < e \quad \text{for all } m, n \geqslant n_e.$$

*(A similar assertion holds for* $(y_n)$.*)*

PROOF: This follows immediately from Theorem 4.5 and the preceding theorem.

In preparation for our definition of the real numbers, we now study some properties of sequences in ordered fields.

In the following, we let $\langle A, +, \cdot, < \rangle$ be an ordered field.

DEFINITION 5.2 A sequence $(a_n)$ in $A$ is *fundamental* if for each $e > 0$ in $A$ there is some $n_e \in \mathbf{N}$ such that

$$|a_n - a_m| < e \text{ in } A \text{ for all } m, n \geqq n_e \text{ in } \mathbf{N}.$$

DEFINITION 5.3 A sequence $(a_n)$ in $A$ is *bounded* if there is some $a \in A$ such that

$$|a_n| < a \text{ in } A \text{ for all } n \in \mathbf{N}.$$

Note that the sequences approximating the gap in Theorem 5.2 are fundamental sequences in $\mathbf{Q}$, by Theorem 4.5.

THEOREM 5.3 *If $(a_n)$ is a fundamental sequence in $A$, then $(a_n)$ is a bounded sequence.*

PROOF: Let $e$ be any positive element of $A$. Then, by Definition 5.2, there is some $n_e \in \mathbf{N}$ such that

(1) $|a_n - a_m| < e$ in $A$ for all $m, n \geqq n_e$ in $\mathbf{N}$.

The finite subset $\{|a_1|, \ldots, |a_{n_e}|\}$ of $A$ contains a maximum element $b$ (Exercise 1, 6.12). Hence,

(2) $|a_n| \leqq b$ in $A$ for $n \leqq n_e$ in $\mathbf{N}$,

and, by (1),

(3) $|a_n| \leqq |a_n - a_{n_e}| + |a_{n_e}| < e + |a_{n_e}|$ in $A$

for all $n > n_e$ in $\mathbf{N}$.

It follows from (2) and (3), by trichotomy, that

$$|a_n| \leqq e + b \text{ in } A$$

for all $n \in \mathbf{N}$.

THEOREM 5.4 *If $(a_n)$and $(b_n)$ are fundamental sequences in $A$, then $(a_n + b_n)$ and $(a_nb_n)$ are fundamental sequences.*

PROOF: Let $e$ be any positive element in $A$. Then $e/2 > 0$ in $A$ and, by hypothesis, there are $n_e', n_e'' \in \mathbf{N}$ such that

(1) $|a_n - a_m| < e/2$ in $A$ for all $n, m \geqq n_e'$ in $\mathbf{N}$,

(2) $|b_n - b_m| < e/2$ in $A$ for all $n, m \geqq n_e''$ in $\mathbf{N}$.

Let $n_e = \max\{n_e', n_e''\}$ in $\mathbf{N}$. Then, by (1) and (2),

$$|(a_n + b_n) - (a_m + b_m)| \leqq |a_n - a_m| + |b_n - b_m| < e/2 + e/2 = e$$

for all $n, m \geqq n_e$ in $\mathbf{N}$.

Hence, $(a_n + b_n)$ is a fundamental sequence in $A$.

By Theorem 5.3, $(a_n)$, $(b_n)$ are bounded sequences, so there are positive elements $a$ and $b$ in $A$ such that

(3) $|a_n| < a$ and $|b_n| < b$ in $A$ for all $n \in \mathbf{N}$.

Since $e/2b$ and $e/2a$ are positive in $A$, there are $n'_e, n''_e \in \mathbf{N}$ such that

(4) $|a_n - a_m| < e/2b$ in $A$ for all $m, n \geqq n'_e$ in $\mathbf{N}$

and

(5) $|b_n - b_m| < e/2a$ in $A$ for all $m, n \geqq n''_e$ in $\mathbf{N}$.

Let $n_e = \max\{n'_e, n''_e\}$ in $\mathbf{N}$. By (3), (4), and (5),

$$|a_n b_n - a_m b_m| \leqq |a_n| \cdot |b_n - b_m| + |b_m| \cdot |a_n - a_m|$$
$$< a \cdot e/2a + b \cdot e/2b = e \text{ in } A \text{ for all } m, n \geqq n_e \text{ in } \mathbf{N}.$$

Hence, $(a_n b_n)$ is a fundamental sequence in $A$.

DEFINITION 5.4 A sequence $(a_n)$ *converges* to $a$ in $A$ if, for each $e > 0$ in $A$, there is some $n_e \in \mathbf{N}$ such that

$$|a_n - a| < e \text{ in } A \text{ for all } n \geqq n_e \text{ in } \mathbf{N}.$$

We call $a$ a *limit* of $(a_n)$ in $A$.

THEOREM 5.5 *A sequence $(a_n)$ has at most one limit in $A$.*

PROOF: Suppose $a'$ and $a''$ are both limits of $(a_n)$. Let $e$ be any positive element in $A$. Then $e/2 > 0$ in $A$ and there are $n'_e, n''_e$ in $\mathbf{N}$ such that

$$|a_n - a'| < e/2 \text{ in } A \text{ for all } n \geqq n'_e \text{ in } \mathbf{N},$$

and

$$|a_n - a''| < e/2 \text{ in } A \text{ for all } n \geqq n''_e \text{ in } \mathbf{N}.$$

Let $n_e = \max\{n'_e, n''_e\}$ in $\mathbf{N}$. Then

$$|a' - a''| \leqq |a' - a_n| + |a_n - a''| < e \text{ in } A \text{ for all } n \geqq n_e.$$

Since $0 \leqq |a' - a''| < e$ in $A$ for all $e > 0$, $a' = a''$.

*Notation:* If $(a_n)$ is convergent, we denote its limit by "$L(a_n)$."

THEOREM 5.6 *If $(a_n)$ is a convergent sequence in $A$, then $(a_n)$ is a fundamental sequence.*

PROOF: Let $a = L(a_n)$ and let $e$ be any positive element in $A$. Then $e/2 > 0$ in $A$, and, since $a$ is the limit of $(a_n)$, there is some $n_e \in \mathbf{N}$ such that

$$|a_n - a| < e/2 \text{ in } A \text{ for all } n \geqq n_e \text{ in } \mathbf{N}.$$

Hence,

$$|a_n - a_m| \leqq |a_n - a| + |a - a_m| < e \text{ in } A$$
$$\text{for all } m, n \geqq n_e \text{ in } \mathbf{N}.$$

THEOREM 5.7 *If $L(a_n) = a$ and $L(b_n) = b$ in $A$, then*

(1) $L(a_n + b_n) = a + b$

*and*

(2) $L(a_n b_n) = ab$.

PROOF: Let $e$ be any positive element in $A$. By hypothesis, there are $n'_e$, $n''_e$ in $\mathbf{N}$ such that

$$(3) \quad |a_n - a| < e/2 \text{ in } A \text{ for all } n \geqq n'_e \text{ in } \mathbf{N},$$
$$|b_n - b| < e/2 \text{ in } A \text{ for all } n \geqq n''_e \text{ in } \mathbf{N}.$$

Let $n_e = \max\{n'_e, n''_e\}$ in $\mathbf{N}$. Then, by (3),

$$|(a_n + b_n) - (a + b)| \leqq |a_n - a| + |b_n - b| < e \text{ in } A$$
$$\text{for all } n \geqq n_e \text{ in } \mathbf{N}.$$

This proves (1).

By Exercise 5.6 (1), there are $a'$, $b'$ in $A$ such that

$$|a_n| < a' \text{ and } |b_n| < b' \text{ for all } n \in \mathbf{N}.$$

Let $c = \max\{a', b',\}$ in $A$. Then $e/2c > 0$ in $A$ and, by hypothesis, there are $\bar{n}_e, \bar{\bar{n}}_e$ in $\mathbf{N}$ such that

$$(4) \quad |a_n - a| < e/2c \text{ in } A \text{ for all } n \geqq \bar{n}_e \text{ in } \mathbf{N},$$
$$|b_n - b| < e/2c \text{ in } A \text{ for all } n \geqq \bar{\bar{n}}_e \text{ in } \mathbf{N}.$$

Let $\tilde{n}_e = \max\{\bar{n}_e, \bar{\bar{n}}_e\}$ in $\mathbf{N}$. Then, by (4) and Exercise 3.18 (2),

$$|a_n b_n - ab| \leqq |a_n||b_n - b| + |b||a_n - a| < a' \frac{e}{2c} \leqq e \text{ in } A$$
$$\text{for all } n \geqq \tilde{n}_e \text{ in } \mathbf{N}.$$

This proves (2).

THEOREM 5.8 *There are non-convergent fundamental sequences in* $\mathbf{Q}$.

PROOF: Let $(X, Y)$ be the gap (associated with "the missing $\sqrt{2}$") of Theorem 5.1. Let $(x_n)$, $(y_n)$ be defined as in Theorem 5.2 for the gap $(X, Y)$. We recall that $y_n = x_n + 1/n < y + 1$ for all $n \in \mathbf{N}$, all $y \in Y$. By Corollary, Th. 5.2, $(x_n)$ is a fundamental sequence in $\mathbf{Q}$.

But $(x_n)$ is not convergent in $\mathbf{Q}$. Otherwise $L(x_n) = z \in \mathbf{Q}$ and, by Theorem 5.7, $L(x_n^2) = z^2$. Now, for each $n \in \mathbf{N}$,

$$0 < 2 - x_n^2 < y_n^2 - x_n^2 = (y_n - x_n)(y_n + x_n) < \frac{2y_n}{n} < \frac{2(y+1)}{n}.$$

Since $L(1/n) = 0$ in $\mathbf{Q}$ (Exercise 5.4), $L(x_n^2) = 2$ in $\mathbf{Q}$. Hence, by Theorem 5.5, $x^2 = 2$ for $x \in \mathbf{Q}$. This is impossible by Theorem 5.1.

THEOREM 5.9 *If $(a_n)$ is a fundamental sequence which does not have limit zero in $A$, then there is a fundamental sequence $(b_n)$ in $A$ such that $L(a_n b_n) = 1$.*

PROOF: Since $(a_n)$ does not have limit zero in $A$, there is a positive element $\bar{e}$ in $A$ such that for every $n \in \mathbf{N}$,

(1) $|a_k| \geqq \bar{e}$ in $A$ for some $k \geqq n$ in $\mathbf{N}$.

Since $(a_n)$ is a fundamental sequence, there is an $\bar{n} \in \mathbf{N}$ such that

(2) $|a_m - a_n| < \frac{\bar{e}}{2}$ in $A$ for all $m, n \geqq \bar{n}$ in $\mathbf{N}$.

If, for $\bar{k} \geqq \bar{n}$,

(3) $|a_{\bar{k}}| \geqq \bar{e}$,

then

(4) $|a_n| = |a_{\bar{k}} - (a_{\bar{k}} - a_n)| \geqq |a_{\bar{k}} - a_n| > \bar{e} - \frac{\bar{e}}{2} = \frac{\bar{e}}{2}$

for all $n \geqq \bar{n}$. Hence, $a_n \neq 0$ for all $n \geqq \bar{n}$.

Now let

(5) $b_1 = 1$ for $n < \bar{n}$
$b_n = 1/a_n$ for $n \geqq \bar{n}$.

Then $(b_n)$ is a sequence in $A$. If $e$ is any positive element of $A$, then there is some $n_e \in \mathbf{N}$ such that

$$|a_m - a_n| < \frac{\bar{e}^2 e}{4}$$

for all $m, n \geqq n_e$. Hence, by (4) and (5),

$$|b_m - b_n| = \frac{|a_m - a_n|}{|a_m||a_n|} < \frac{\bar{e}^2 e}{4} \cdot \frac{4}{\bar{e}^2} = e$$

for all $m, n \geqq \max(\bar{n}, n_e)$. It follows that $(b_n)$ is a fundamental sequence in $A$.

Since $a_n b_n = 1$ for all $n \geqq \bar{n}$, $L(a_n b_n) = 1$.

DEFINITION 5.5 A sequence $(a_n)$ in $A$ is *positive* if, for some positive $e$ in $A$ and some $k \in \mathbf{N}$,

$$a_n \geqq e \text{ in } A \text{ for all } n \geqq k \text{ in } \mathbf{N}.$$

*Example:* The sequence $(1 + \frac{1}{n})$ in $\mathbf{Q}$ is positive, but the sequence $(\frac{1}{n})$ is not. (See Exercise 5.9.)

THEOREM 5.10 *If $(a_n)$ is a fundamental sequence in $A$, then exactly one of the following statements is true:*

(1) $L(a_n) = 0$.
(2) *$(a_n)$ is positive.*
(3) *$(-a_n)$ is positive.*

PROOF: We first show that at least one of (1), (2), and (3) is true. Suppose (1) is false. Then there is a positive element $e$ in $A$ such that for each $m \in \mathbf{N}$

$$(4) \quad |a_k| \geqq e \text{ in } A \text{ for some } k \geqq m \text{ in } \mathbf{N}.$$

Since $e/2 > 0$ in $A$ and $(a_n)$ is a fundamental sequence, there is some $n_e \in \mathbf{N}$ such that

$$(5) \quad |a_n - a_m| < e/2 \text{ in } A \text{ for all } m, n \geqq n_e \text{ in } \mathbf{N}.$$

By (4), with $m = n_e$,

$$(6) \quad \max\{a_k, -a_k\} = |a_k| \geqq e \text{ in } A \text{ for some } k \geqq n_e \text{ in } \mathbf{N}.$$

If $a_k \geqq e$ then, by (5),

$$a_n = a_k - (a_k - a_n) \geqq e - |a_k - a_n| > e/2 \text{ in } A$$
$$\text{for all } n \geqq n_e \text{ in } \mathbf{N}.$$

Hence, by Definition 5.5, $(a_n)$ is positive and (2) is true. Otherwise, by (6), $-a_k \geqq e$ and, again by (5),

$$-a_n = -a_k - (a_n - a_k) \geqq e - |a_n - a_k| > e/2 \text{ in } A$$
$$\text{for all } n \geqq n_e \text{ in } \mathbf{N}.$$

Hence $(-a_n)$ is positive and (3) is true.

Next we show that not more than one of the three statements is true. If $L(a_n) = 0$, then for each $e > 0$ in $A$ there is some $n_e$ in $\mathbf{N}$ such that

$$\max\{a_n, -a_n\} = |a_n| < e \text{ in } A \text{ for all } n \geqq n_e \text{ in } \mathbf{N}.$$

Hence, there is no positive $e$ in $A$ such that, for some $k \in \mathbf{N}$, either

$$a_n \geqq e \text{ in } A \text{ for all } n \geqq k \text{ in } \mathbf{N}$$

or

$$-a_n \geqq e \text{ in } A \text{ for all } n \geqq k \text{ in } \mathbf{N}.$$

Thus if (1) is true, then (2) and (3) are both false. If (2) and (3) are both true, then for some $e'$ and $e''$, positive in $A$, and $k', k'' \in \mathbf{N}$.

$$a_n \geqq e' \text{ in } A \text{ for all } n \geqq k' \text{ in } \mathbf{N}$$

and

$$-a_n \geqq e'' \text{ in } A \text{ for all } n \geqq k'' \text{ in } \mathbf{N}.$$

Hence, for $n = \max\{k', k''\}$ in $\mathbf{N}$,

$$0 < e'' \leqq -a_n \leqq -e' < 0 \text{ in } A.$$

But this is impossible.

It follows that exactly one of (1), (2), and (3) must hold.

• *Exercise 5.1* Prove Theorem 5.1, 2a, b, c.

*Exercise 5.2* Let $(x_n)$, $(y_n)$ be the sequences defined in Theorem 5.2. For each $n \in \mathbf{N}$, let $J_n = \{x \in \mathbf{Q} | x_n \leqslant x \leqslant y_n\}$. Prove that $\cap J_n = \phi$.

*Exercise 5.3* Prove Theorem 5.3.

*Exercise 5.4* Prove: an ordered field $A$ is Archimedean if and only if $L(\frac{1}{n}) = 0$ in $A$. (See Theorem 4.5.)

*Exercise 5.5* Prove: an ordered field $A$ is Archimedean if and only if $L(\frac{1}{p^n}) = 0$ in $A$ for each $p \neq 1$ in $\mathbf{N}$.

• *Exercise 5.6*

(1) If $(a_n)$ is a convergent sequence in $A$, then $(a_n)$ is a bounded sequence.

(2) If $|a_n| \leqq b$ for all $n \in \mathbf{N}$ and $L(a_n) = a$, then $|a| \leqq b$.

*Exercise 5.7*

(1) For all $a \in A$, if $L(a_n) = a$ in $A$, then $L(|a_n|) = |a|$.

(2) Show that the converse of (1) is false in any ordered field.

(3) $L(|a_n|) = 0$ in $A$ if and only if $L(a_n) = 0$.

*Exercise 5.8* Show that the conditions stated below define a sequence $(a_n)$ which is fundamental but not convergent in $\mathbf{Q}$:

$$a_1 = 1$$

$$a_{n+1} = a_n + \frac{b_n}{10^n},$$

where $b_n$ is a non-negative integer such that

$$\left(a_n + \frac{b_n}{10^n}\right)^2 < 2 < \left(a_n + \frac{b_n + 1}{10^n}\right)^2$$

Note: The $a_n$ are "decimal approximations to $\sqrt{2}$".

• *Exercise 5.9*

(a) Prove that, in any ordered field $A$, the sequence $(1 + \frac{1}{n})$ is positive.

(b) Prove that, in an Archimedean ordered field, the sequence $(\frac{1}{n})$ is not positive.

• *Exercise 5.10* Let $F_A$ be the set of all fundamental sequences in the ordered field $A$. Let $T$ be the subset of $F_A \times F_A$ consisting of those $((a_n), (b_n)) \in F_A \times F_A$ for which $(b_n - a_n) \in F_A$ is positive. Show that $T$ is not an order relation in $F_A$.

*Exercise 5.11* If $(a_n)$ is a fundamental sequence in an ordered field $A$ and $(a_n)$ does not have limit zero, then $(|a_n|)$ is a positive fundamental sequence in $A$.

• *Exercise 5.12* If $(a_n)$ and $(b_n)$ are positive fundamental sequences in $A$, then $(a_n + b_n)$ and $(a_n b_n)$ are positive in $A$.

• *Exercise 5.13* If $A$, $B$ are ordered fields and $F\colon A \to B$ is a bijection which preserves addition, multiplication, and order, then

(1) $(a_n)$ is a fundamental sequence in $A$ if and only if $(F(a_n))$ is a fundamental sequence in $B$;
(2) $L(a_n) = a$ if and only if $L(F(a_n)) = F(a)$;
(3) $(a_n)$ is a positive sequence in $A$ if and only if $(F(a_n))$ is a positive sequence in $B$.

• *Exercise 5.14* If $(x_n)$ is a convergent sequence in an ordered field $A$, then $(x_n)$ is a positive sequence in $A$ if and only if $L(x_n) > 0$ in $A$.

# CHAPTER 4

# THE REAL NUMBERS

**1. THE FIELD OF REAL NUMBERS.** We introduced the integers as equivalence classes of ordered pairs of natural numbers, and the rational numbers as equivalence classes of ordered pairs of integers. In the construction of the real numbers we again begin with the definition of an equivalence relation. In this case, the equivalence relation will be defined in the set of all fundamental rational sequences.

Thus, a real number will be an equivalence class of fundamental rational sequences. By suitable definitions of addition, multiplication, and order, the set **R** of all real numbers will be made into an ordered field which will be an extension of the ordered field **Q**. The order in **R** will have no gaps in the sense of Definition 3, 5.1. Equivalently, every fundamental sequence of real numbers will have a limit in **R**, i.e., the converse of Theorem 3, 5.6 (which is false in **Q**, by Theorem 3, 5.8) will hold in **R**.

We shall use "$F_{\mathrm{Q}}$" to denote the set of all fundamental rational sequences.

THEOREM 1.1 *There is an equivalence relation $Q$ in $F_{\mathrm{Q}}$ such that $(x_n)Q(y_n)$ holds whenever $L(x_n - y_n) = 0$.*

PROOF: The set

$$Q = \{((x_n), (y_n)) \mid L(x_n - y_n) = 0\}$$

is a subset of $F_{\mathrm{Q}} \times F_{\mathrm{Q}}$. Since $L(x_n - x_n) = L(0) = 0$ for each $(x_n) \in F_{\mathrm{Q}}$, $Q$ is reflexive. If $L(x_n - y_n) = 0$, then $L(-[x_n - y_n]) = L(y_n - x_n) = 0$, so that $Q$ is symmetric.

If $L(x_n - y_n) = 0$ and $L(y_n - z_n) = 0$, then $L(x_n - z_n) = L(x_n - y_n + y_n - z_n) = L([x_n - y_n] + [y_n - z_n]) = L(x_n - y_n) + L(y_n - z_n) = 0$. Hence, $Q$ is transitive.

As usual, we write $(x_n) \sim (y_n)$ if the pair $((x_n), (y_n)) \in Q$, and denote by "$C_{(x_n)}$" the equivalence class containing $(x_n)$.

DEFINITION 1.1 A *real number* is an equivalence class $C_{(x_n)}$ with respect to the equivalence relation $Q$ of Theorem 1.1, where $(x_n)$ is a fundamental rational sequence.

We denote by $\mathbf{R}$ the set of all real numbers and note that $\mathbf{R}$ is the factor set $F_{\mathbf{Q}}/Q$. We use $\xi, \eta, \ldots$ to denote real numbers.

THEOREM 1.2 *If* $(x_n), (y_n), (x'_n), (y'_n) \in F_{\mathbf{Q}}$, $(x_n) \sim (x'_n)$ *and* $(y_n) \sim (y'_n)$, *then*

(1) $(x_n + y_n) \sim (x'_n + y'_n)$

*and*

(2) $(x_n y_n) \sim (x'_n y'_n)$.

PROOF:

(1) By Theorem 3, 5.7, $L([x_n + y_n] - [x'_n + y'_n]) = L(x_n - x'_n + y_n - y'_n) = L(x_n - x'_n) + L(y_n - y'_n) = 0 + 0 = 0$ in $\mathbf{Q}$.

(2) By Theorem 3, 5.3, since $(x_n), (y'_n) \in F_{\mathbf{Q}}$, there exist $a, b \in \mathbf{Q}$ such that $|x_n| < a$, $|y'_n| < b$ for all $n \in \mathbf{N}$.

By Theorem 3, 5.4, $(x_n y_n)$ and $(x'_n y'_n)$ are fundamental sequences in $\mathbf{Q}$. Since $(x_n) \sim (x'_n)$ and $(y_n) \sim (y'_n)$, there are, for each positive $e$ in $\mathbf{Q}$, $n'_e$ and $n''_e$ in $\mathbf{N}$ such that

$$|x_n - x'_n| < \frac{e}{2b} \text{ in } \mathbf{Q} \text{ for all } n \geqq n'_e \text{ in } \mathbf{N}$$

and

$$|y_n - y'_n| < \frac{e}{2a} \text{ in } \mathbf{Q} \text{ for all } n \geqq n''_e \text{ in } \mathbf{N}.$$

Hence,

$$\begin{aligned} |x_n y_n - x'_n y'_n| &\leqq |x_n||y_n - y'_n| + |y'_n||x_n - x'_n| \\ &< a \cdot \frac{e}{2a} + b \cdot \frac{e}{2b} = e \end{aligned}$$

$$\text{for all } n \geqq n_e = \max\{n'_e, n''_e\} \text{ in } \mathbf{N}.$$

Therefore $L(x_n y_n - x'_n y'_n) = 0$ in $\mathbf{Q}$ and $(x_n y_n) \sim (x'_n y'_n)$.

THEOREM 1.3 *There are binary operations $F$ and $G$ on $\mathbf{R}$ such that, if $(x_n) \in \xi$ and $(y_n) \in \eta$, then*

(1) $F(\xi, \eta) = C_{(x_n + y_n)}$

(2) $G(\xi, \eta) = C_{(x_n y_n)}$.

PROOF: The sets

$$F = \{((\xi, \eta), C_{(x_n + y_n)}) \mid (x_n) \in \xi, (y_n) \in \eta; \xi, \eta \in \mathbf{R}\}$$

and

$$G = \{((\xi, \eta), C_{(x_n y_n)}) \mid (x_n) \in \xi, (y_n) \in \eta; \xi, \eta \in \mathbf{R}\}$$

are subsets of $(\mathbf{R} \times \mathbf{R}) \times \mathbf{R}$.

If $(\xi, \eta) \in \mathbf{R} \times \mathbf{R}$, then $\xi = C_{(x_n)}$, $\eta = C_{(y_n)}$for some $(x_n)$, $(y_n) \in F_\mathbf{Q}$. Since $(x_n + y_n) \in F_\mathbf{Q}$ by Theorem 3, 5.4, the pair $((\xi, \eta), \zeta) \in F$ where $\zeta = C_{(x_n + y_n)}$. If $((\xi, \eta), \zeta') \in F$, then $\zeta' = C_{(x'_n + y'_n)}$ where $(x'_n) \in \xi$, $(y'_n) \in \eta$. By Theorem 4.2, since $(x'_n) \sim (x_n)$, $(y'_n) \sim (y_n)$, it follows that $(x'_n + y'_n) \sim (x_n + y_n)$ and $\zeta = \zeta'$. Thus $F$ is a mapping of $(\mathbf{R} \times \mathbf{R})$ into $\mathbf{R}$, and hence a binary operation on $\mathbf{R}$.

If $(\xi, \eta) \in \mathbf{R} \times \mathbf{R}$, then $(x_n) \in \xi$ and $(y_n) \in \eta$ for some $(x_n)$, $(y_n) \in F_\mathbf{Q}$. By Theorem 3, 5.4, $(x_n y_n) \in F_\mathbf{Q}$. Hence, $((\xi, \eta), \zeta) \in G$ where $\zeta = C_{(x_n y_n)}$. If $((\xi, \eta), \zeta') \in G$, then $\zeta' = C_{(x'_n y'_n)}$ where $(x'_n) \in \xi$, $(y'_n) \in \eta$. By Theorem 3, 5.4, since $(x'_n) \sim (x_n)$, $(y'_n) \sim (y_n)$, it follows that $(x_n y_n) \sim (x'_n y'_n)$, and $\zeta = \zeta'$. Thus $G$ is a mapping of $\mathbf{R} \times \mathbf{R}$ into $\mathbf{R}$, and hence a binary operation on $\mathbf{R}$.

DEFINITION 1.2 We call the binary operations $F$ and $G$ of Theorem 1.3 *addition in* $\mathbf{R}$ and *multiplication in* $\mathbf{R}$, respectively, and write "$\xi +_\mathbf{R} \eta$" and "$\xi \cdot_\mathbf{R} \eta$" for $F(\xi, \eta)$ and $G(\xi, \eta)$. As usual, we shall feel free to omit the subscript "$\mathbf{R}$."

THEOREM 1.4 *$\langle \mathbf{R}, +_\mathbf{R}, \cdot_\mathbf{R} \rangle$ is a field.*

PROOF: We leave to the reader the verification of the associative, commutative, and distributive properties. We note that $C_{(0)}$ serves as the additive identity, $0_\mathbf{R}$; $C_{(1)}$ as the multiplicative identity, $1_\mathbf{R}$; and $C_{(-x_n)}$ as the additive inverse $-C_{(x_n)}$ of $C_{(x_n)}$. If $C_{(x_n)} \neq 0_\mathbf{R}$, then $(x_n)$ is not equivalent to $(0)$, so that $(x_n)$ does not have limit zero in $\mathbf{Q}$. By Theorem 3, 5.9, there is a sequence $(y_n) \in F_\mathbf{Q}$ such that $L(x_n y_n) = 1_\mathbf{Q}$. Hence, $(x_n y_n) \sim (1)$, and $C_{(x_n)} C_{(y_n)} = C_{(x_n y_n)} = C_{(1)} = 1_\mathbf{R}$. Thus, $C_{(y_n)}$ is the multiplicative inverse, $\dfrac{1}{C_{(x_n)}}$, of $C_{(x_n)}$. It follows that $\langle \mathbf{R}, +_\mathbf{R}, \cdot_\mathbf{R} \rangle$ is a field.

*Remark:* $0_\mathbf{R} = C_{(0)} = \{(x_n) \in F_\mathbf{Q} \mid L(x_n) = 0\}$, and
$1_\mathbf{R} = C_{(1)} = \{(x_n) \in F_\mathbf{Q} \mid L(x_n) = 1\}$.
(See Exercise 2.1).

• *Exercise 2.1* Prove that $0_\mathbf{R} = \{(x_n) \in F_\mathbf{Q} \mid L(x_n) = 0\}$ and
$1_\mathbf{R} = \{(x_n) \in F_\mathbf{Q} \mid L(x_n) = 1\}$.

**2. ORDER IN R.** We shall define the positive elements of $\mathbf{R}$ as the equivalence classes belonging to positive sequences in $F_{\mathbf{R}}$.

*Notation:* We let $\mathbf{R}^+ = \{\xi \in \mathbf{R} \mid \text{for some } (x_n) \in \xi, (x_n) \text{ is positive}\}$.

THEOREM 2.1 *If $(x_n) \sim (x'_n)$, and $(x_n)$ is a positive sequence, then $(x'_n)$ is a positive sequence.*

PROOF: If $(x_n)$ is a positive sequence, then by Definition 3, 5.5, there are $e > 0$ in $\mathbf{Q}$ and $n_e \in \mathbf{N}$ such that $x_n \geqq e$ for $n \geqq n_e$. Since $(x_n) \sim (x'_n)$, there is some $n'_e$ in $\mathbf{N}$ such that

$$|x'_n - x_n| < \frac{e}{2} \text{ for } n \geqq n'_e \text{ in } \mathbf{N}.$$

Hence, $-\frac{e}{2} < x'_n - x_n < \frac{e}{2}$ for $n \geqq n'_e$. But then, if $\bar{n}_e = \max\{n_e, n'_e\}$ in $\mathbf{N}$,

$$x'_n = (x'_n - x_n) + x_n > -\frac{e}{2} + e - \frac{e}{2} > 0 \text{ for all } n \geqq \bar{n}_e.$$

Hence $(x'_n)$ is a positive sequence in $\mathbf{Q}$.

COROLLARY $\mathbf{R}^+ = \{\xi \in \mathbf{R} \mid (x_n) \text{ is positive for all } (x_n) \in \xi\}$.

THEOREM 2.2 *$\mathbf{R}^+$ is a set of positive elements for $\mathbf{R}$.*

PROOF: We show that $\mathbf{R}^+$ satisfies (1), (2), and (3) of Definition 2, 7.3. If $\xi, \eta \in \mathbf{R}^+$, then $\xi = C_{(x_n)}$, $\eta = C_{(y_n)}$, where $(x_n), (y_n)$ are positive sequences in $\mathbf{Q}$. By Exercise 3, 5.13, $\xi + \eta = C_{(x_n + y_n)} \in \mathbf{R}^+$, and $\xi\eta = C_{(x_n y_n)} \in \mathbf{R}^+$, so that (1) and (2) are fulfilled.

If $\xi = C_{(x_n)}$, then, by the Corollary of Theorem 2.1,

$$\xi \in \mathbf{R}^+ \text{ if and only if } (x_n) \text{ is a positive sequence in } \mathbf{Q},$$
$$\xi = 0_{\mathbf{R}} = C_{(0)} \text{ if and only if } L(x_n) = 0_{\mathbf{Q}},$$
$$-\xi = C_{(-x_n)} \in \mathbf{R}^+ \text{ if and only if } (-x_n) \text{ is a positive sequence in } \mathbf{Q}.$$

Hence, by Theorem 3, 5.10, exactly one of

$$\xi \in \mathbf{R}^+, \qquad \xi = 0_{\mathbf{R}}, \qquad -\xi \in \mathbf{R}^+$$

must hold. Thus, (3) is fulfilled, and $\mathbf{R}^+$ is a set of positive elements for $\mathbf{R}$.

By Theorem 2, 7.2 (1), we have

THEOREM 2.3 *The set $T = \{(\xi, \eta) \mid \eta - \xi \in \mathbf{R}^+\}$ is an order relation in $\mathbf{R}$.*

*Notation:* We write "$\xi <_{\mathbf{R}} \eta$" ("$\eta >_{\mathbf{R}} \xi$") if $(\xi, \eta) \in T$. Usually, we omit the subscript "$\mathbf{R}$."

By Theorem 2.4 (2), and Definition 3, 2.1, we have

THEOREM 2.4 *$\langle \mathbf{R}, +_{\mathbf{R}}, \cdot_{\mathbf{R}}, <_{\mathbf{R}} \rangle$ is an ordered field.*

## 3. EMBEDDING.

THEOREM 3.1 *If $E$: $\mathbf{Q} \to \mathbf{R}$ is defined by $E(x) = C_{(x)}$ for each $x \in \mathbf{Q}$, then $E$ is an isomorphic embedding of* $\mathbf{Q}$ into $\mathbf{R}$ *with respect to addition, multiplication and order.*

PROOF: $E$ is a bijection from $\mathbf{Q}$ to $\mathbf{R}$.
For, $C_{(x)} = C_{(y)}$ if and only if $(x) \sim (y)$, i.e., if and only if $x = y$ in $\mathbf{Q}$. If $x, y \in \mathbf{Q}$, then

$$E(x+y) = C_{(x+y)} = C_{(x)} +_{\mathbf{R}} C_{(y)} = E(x) +_{\mathbf{R}} E(y),$$

and

$$E(xy) = C_{(xy)} = C_{(x)} \cdot_{\mathbf{R}} C_{(y)} = E(x) \cdot_{\mathbf{R}} E(y).$$

Thus, $E$ preserves addition and multiplication.

Also, $x < y$ in $\mathbf{Q}$ if and only if $y - x > 0$ in $\mathbf{Q}$ so that $(y - x)$ is a positive sequence in $F_{\mathbf{Q}}$. On the other hand, $C_{(x)} < C_{(y)}$ in $\mathbf{R}$ if and only if $C_{(y)} - C_{(x)} > 0$ in $\mathbf{R}$, so that $(y - x)$ is a positive sequence in $F_{\mathbf{Q}}$. Thus, $x < y$ in $\mathbf{Q}$ if and only if $C_{(x)} <_{\mathbf{R}} C_{(y)}$ in $\mathbf{R}$, and $E$ preserves order.

As usual, we shall identify $\mathbf{Q}$ with its isomorphic image in $\mathbf{R}$ and use interchangeably the symbols $x$ and $C_{(x)}$.

THEOREM 3.2 *For every real number $\varepsilon > 0$, there is a rational number $e$ such that $0 < e < \varepsilon$.*

PROOF: Let $(x_n)$ be a fundamental rational sequence such that $\varepsilon = C_{(x_n)}$. Since $\varepsilon > 0$ in $\mathbf{R}$, $(x_n)$ is a positive sequence. Hence, for some $\bar{n} \in \mathbf{N}$ and some $\bar{e} \in Q^+$, $x_n \geqslant e$ when $n \geqslant \bar{n}$. Since $x_n - \bar{e}/2 \geqslant \bar{e}/2 > 0$, $(x_n - \bar{e}/2)$ is a positive fundamental sequence, and so $C_{(x_n - \bar{e}/2)} > 0$ in $\mathbf{R}$. But then $C_{(x_n)} - C_{(\bar{e}/2)} > 0$ in $\mathbf{R}$, and $0 < C_{(\bar{e}/2)} < C_{(x_n)} = \varepsilon$. Setting $e = C_{(\bar{e}/2)}$, we have $0 < e < \varepsilon$, with $e \in \mathbf{Q}$.

COROLLARY *A rational sequence $(x_n)$ is fundamental in* $\mathbf{Q}$ *if and only if it is fundamental in* $\mathbf{R}$.

PROOF: See Exercise 3.2.

*Exercise 3.1* For each $x \in \mathbf{Q}, E(x) = \{(x_n) \in F_{\mathbf{Q}} \mid L(x_n) = x\}$.

• *Exercise 3.2* Prove that a rational sequence $(x_n)$ is fundamental in $\mathbf{Q}$ if and only if it is fundamental in $\mathbf{R}$.

• *Exercise 3.3* For $(x_n) \in F_Q$, $|C_{(x_n)}| = C_{(|x_n|)}$.

4. **COMPLETENESS OF R.** We have shown (Theorem 3, 5.6) that in an ordered field every convergent sequence is fundamental and (Theorem 3, 5.8) that the converse does not hold in the rational field.

DEFINITION 4.1 An ordered field $A$ is called *complete* if every fundamental sequence in $A$ is convergent.

We shall show that **R** is complete. We first prove that every fundamental rational sequence converges in **R**.

THEOREM 4.1 *If* $(x_n) \in \xi$, *then* $L(x_n) = \xi$ *in* **R**.

PROOF: For $e > 0$ in **R**, let $e$ be a rational number such that $0 < e < \varepsilon$. Since $(x_n)$ is fundamental in **Q**, there is an $n_e$ in **N** such that

$$|x_m - x_n| < \frac{e}{2} \text{ for all } m, n \geqq n_e.$$

Hence, for all $m, n \geqq n_e$,

$$e - |x_n - x_m| > \frac{e}{2}$$

and for each $n \geqq n_e$, $(y_m) = (e - |x_n - x_m|)$ is a positive fundamental sequence in **R**. It follows that, for each $n \geqq n_e$,

$$C_{(y_m)} = C_{(e - |x_n - x_m|)} > 0 \text{ in } \mathbf{R}.$$

But then

$$|x_n - \xi| = |x_n - C_{(x_m)}| = C_{(|x_n - x_m|)} < C_{(e)} = e < \varepsilon \text{ for all } n \geqq n_e,$$

and

$$L(x_n) = \xi \text{ in } \mathbf{R}.$$

COROLLARY 1 *If* $\xi \in \mathbf{R}$ *and* $\varepsilon > 0$ *in* **R**, *there is an* $x \in \mathbf{Q}$ *such that* $|\xi - x| < \varepsilon$ *in* **R**.

PROOF: If $(x_n) \in \xi$, then $\xi = L(x_n)$. Hence, for every $e > 0$ in **R**, there is an $n_\varepsilon$ in **N** such that

$$|\xi - x_n| < \varepsilon \text{ in } \mathbf{R} \text{ for all } n \geqq n_\varepsilon.$$

In particular, for $x = x_{n_\varepsilon} \in \mathbf{Q}$, $|\xi - x| < \varepsilon$, as required.

COROLLARY 2 *If* $\xi < \eta$ *in* **R**, *there is a* $z \in \mathbf{Q}$ *such that* $\xi < z < \eta$.

PROOF: By Theorem 3, 4.3, there is a real number $\zeta$ such that $\xi < \zeta < \eta$. If $\varepsilon = \min\{\zeta - \xi, \eta - \zeta\}$, then, by Corollary 1, there is a rational number $z$ such that $\xi \leqq \zeta - \varepsilon < z < \zeta + \varepsilon \leqq \eta$.

COROLLARY 3 **R** *is Archimedean.*

PROOF: For $0 < \xi < \eta$ in **R**, let $x$, $y$ be rational numbers such that

$$0 < x < \xi < \eta \leqq y < \xi + \eta \text{ in } \mathbf{R}.$$

Since **Q** is Archimedean and the embedding preserves addition and order, there is an $n \in \mathbf{N}$ such that $nx \geqq y$ in **R**. But then $n\xi > nx \geqq y \geqq \eta$, and **R** is Archimedean.

THEOREM 4.2 **R** *is complete.*

PROOF: Let $(\xi_n)$ be a fundamental sequence in **R**. By Corollary 1, there is, for each $n \in \mathbf{N}$, a rational number $z_n$ such that

$$|\xi_n - z_n| < \frac{1}{n}.$$

We show that $(z_n)$ is a fundamental sequence in **R**. Since **R** is Archimedean (or also by Exercise 4.1), $L(1/n) = 0$ in **R**. Hence, for every $\varepsilon > 0$ in **R**, there is an $n_1 \in \mathbf{N}$ such that

$$|\xi_n - z_n| < \frac{1}{n} < \frac{\varepsilon}{3} \text{ for all } n \geqq n_1.$$

Since $(\xi_n)$ is a fundamental sequence, there is an $n_2 \in \mathbf{N}$ such that

$$|\xi_m - \xi_n| < \frac{\varepsilon}{3} \text{ for all } m, n \geqq n_2.$$

Hence,

$$|z_m - z_n| \leqq |z_m - \xi_m| + |\xi_m - \xi_n| + |\xi_n - z_n| < \frac{\varepsilon}{3} + \frac{\varepsilon}{3} + \frac{\varepsilon}{3} = \varepsilon$$

for all $m, n \geqq \max\{n_1, n_2\}$. Thus, $(z_n)$ is a fundamental sequence in **Q**. By Theorem 4.1*, $L(z_n) = C_{(z_n)} = \xi$ in **R**. Hence, there is an $n_3 \in \mathbf{N}$ such that

$$|z_n - \xi| < \frac{2\varepsilon}{3} \text{ for all } n \geqq n_3.$$

But then

$$|\xi_n - \xi| \leqq |\xi_n - z_n| + |z_n - \xi| < \frac{\varepsilon}{3} + \frac{2\varepsilon}{3} = \varepsilon$$

for all $n \geqq \max\{n_1, n_3\}$. Hence, $L(\xi_n) = \xi$.

DEFINITION 4.2 If $B$ is a subset of an ordered set $A$, then $B$ is *dense in* $A$ if for all $a$, $b$ in $A$ such that $a < b$ in $A$, there is some element $c \in B$ such that $a < c < b$.

By Corollary 2 of Theorem 4.1, **Q** is dense in **R**. The close connection between the "density" of **Q** in **R** and the Archimedean property of **R** is expressed in the following more general theorem:

THEOREM 4.3 *An ordered field $\langle A, +, \cdot, < \rangle$ is Archimedean if and only if the subset $\mathbf{Q}_A$ of all rational elements of $A$ is dense in $A$.*

PROOF: Suppose $A$ is Archimedean. For $a < b$ in $A$, one of the following holds:

(1) $a = 0 < b$,
(2) $a < b = 0$,
(3) $a < 0 < b$,
(4) $0 < a < b$,
(5) $a < b < 0$.

By Exercise 3, 5.6, $L(1/n) = 0$ in $A$. Hence if (1) holds, then $a = 0 < 1/n < b$ for some $n \in \mathbf{N}$, and if (2) holds, then $a < -1/n < b = 0$ for some $n \in \mathbf{N}$. If (3) holds, then 0 is a rational element between $a$ and $b$. If (4) holds, then there is some $n \in \mathbf{N}$ such that

(1) $0 < 1/n < b - a$,

and there is a least $m \in \mathbf{N}$ such that

(2) $b \leqq \dfrac{m}{n}$.

Hence $(m-1)/n < b$. Also,

(3) $$\frac{m-1}{n} = \frac{m}{n} - \frac{1}{n} > b - (b - a) = a.$$

But then

(4) $$a < \frac{m-1}{n} < b.$$

Finally, if (5) holds, then $0 < -b < -a$, and, by (4), there is a rational element $r$ such that

$$-b < r < -a.$$

But then

$$a < -r < b.$$

It follows that $\mathbf{Q}_A$ is dense in $A$ (Definition 4.2).

Now suppose $\mathbf{Q}_A$ is dense in $A$. Then, for $0 < a < b$ in $A$, there are rational elements $x$ and $y$ such that

$$0 < x < a < b < y < a + b.$$

Since $\mathbf{Q}_A$ is Archimedean, and the isomorphic embedding (Exercise 3, 4.1) preserves addition and order, there is some $n \in \mathbf{N}$ such that

$$nx \geqq y \text{ in } A.$$

Hence, $na > nx \geqq y > b$, and $A$ is Archimedean.

*Exercise 4.1* If $(x_n)$ is a rational sequence and $x \in \mathbf{Q}$, then

$$L(x_n) = x \text{ in } \mathbf{Q} \text{ if and only if } L(x_n) = x \text{ in } \mathbf{R}.$$

• *Exercise 4.2* If $B$ and $A$ are ordered fields such that $B$ is dense in $A$ and $(a_n)$ is a sequence in $B$, then

(1) $(a_n)$ is a fundamental sequence in $A$ if and only if it is a fundamental sequence in $B$.
(2) $L(a_n) = a$ in $A$ if and only if $L(a_n) = a$ in $B$,
(3) $(a_n)$ is a positive sequence in $A$ if and only if it is a positive sequence in $B$.

• *Exercise 4.3* An ordered field $A$ is Archimedean if and only if every element of $A$ is the limit of a sequence of rational elements of $A$.

*Exercise 4.4* Prove that $\mathbf{R}$ contains a square root of 2. (Hint: Recall Theorem 3, 5.2).

*Exercise 4.5* Justify each step in the following alternate proof of Theorem 4.2:

(1) If $k \in \mathbf{N}$, then there is some $n_k \in \mathbf{N}$ such that $k \leqslant n_k < n_{k+1}$ in $\mathbf{N}$ and $|\xi_n - \xi_{n_k}| < 1/2^k$ in $\mathbf{R}$ for all $k$.
(2) If $k \in \mathbf{N}$, there are $(x_{k,m}) \in \xi_{n_k}$ and $m_k \in \mathbf{N}$ such that $n_k \leqslant m_k < m_{k+1}$ in $\mathbf{N}$ and

$$|x_{k,m} - x_{k,m_k}| < 1/2^k \text{ in } \mathbf{Q} \text{ for all } k \text{ and all } m \geqslant m_k \text{ in } \mathbf{N}.$$

(3) For $k \in \mathbf{N}$, write $x_k = x_{k,m_k}$. Then

$$|x_n - x_k| \leqslant \sum_{j=k}^{h-1} |x_{j+1} - x_j| < 1/2^{k-1} \text{ in } \mathbf{Q} \text{ for all } h > k \text{ in } \mathbf{N}.$$

(4) $(x_k) \in F_{\mathbf{Q}}$ and $\xi = C_{(x_k)} \in \mathbf{R}$.

(5) $|x_{k,m} - x_m| \leqslant |x_{k,m} - x_{k,m_k}| + |x_k - x_m| < 1/2^{k-2}$ in **Q** for all $k$, and all $m \geqslant m_k$ in **N**.
(6) $|\xi_n - \xi| \leqslant |\xi_n - \xi_{n_k}| + |\xi_{n_k} - \xi| < \frac{1}{2^{k-3}}$ in **R** for all $k$ and all $n \geqslant n_k$ in **N**.
(7) $L(1/2^{k-3}) = 0$ in **R**.
(8) $L(\xi_n) = \xi$ in **R**.

5. METRIC SPACES. We now make precise the remark in Chapter 3 that the function $|a - b|$ from $A \times A$ to $A$, where $A$ is an ordered field, plays the role of a distance in $A$.

DEFINITION 5.1 Let $S$ be a set and let **D** be a mapping of $\mathbf{S} \times \mathbf{S}$ into **R** such that

(1) $\mathbf{D}(P, Q) \geqq 0$ in **R** for all $P, Q \in \mathbf{S}$ and the equality holds if and only if $P = Q$. (**D** is non-negative.)
(2) $\mathbf{D}(P, Q) = \mathbf{D}(Q, P)$ for all $P, Q \in \mathbf{S}$. (**D** is symmetric.)
(3) $\mathbf{D}(P, Q) \leqq \mathbf{D}(P, V) + \mathbf{D}(V, Q)$ for all $P, V, Q \in \mathbf{S}$. (**D** satisfies the triangle inequality.)

The system $\langle \mathbf{S}, \mathbf{D} \rangle$ is called a *metric space* **S**, and **D** is called a *distance function (metric) for S*.

*Examples of metric spaces:* $\langle \mathbf{Q}, \tilde{\mathbf{d}} \rangle$, where **Q** is the rational field and $\tilde{\mathbf{d}}(x, y) = |x - y|$.

$\langle \mathbf{R}, \mathbf{d} \rangle$, where **R** is the real field and $\mathbf{d}(\xi, \eta) = |\xi - \eta|$.

$\langle \mathbf{Q}^{(n)}, \tilde{\mathbf{d}}^{(n)} \rangle$, where $\mathbf{Q}^{(n)}$ is the set of all $n$-tuples $x^{(n)} = \langle x_1, \ldots, x_n \rangle$, $x_j \in \mathbf{Q}$, and $\tilde{\mathbf{d}}^{(n)}(x^{(n)}, y^{(n)}) = \left( \sum_{j=1}^{n} (x_j - y_j)^2 \right)^{1/2}$

$\langle \mathbf{R}^{(n)}, \mathbf{d}^{(n)} \rangle$, where $\mathbf{R}^{(n)}$ is the set of all $n$-tuples, $\xi^{(n)} = \langle \xi_1, \ldots, \xi_n \rangle$, $\xi_j \in \mathbf{R}$, and $\mathbf{d}^{(n)}(\xi^{(n)}, \eta^{(n)}) = \left( \sum_{j=1}^{n} (\xi_j - \eta_j)^2 \right)^{1/2}$ (Euclidean space).

$\langle \mathbf{R}^{(n)}, \Delta^{(n)} \rangle$, where $\Delta^{(n)}(\xi^{(n)}, \eta^{(n)}) = \max \{ |\xi_j - \eta_j| \,|\, j = 1, \ldots, n \}$ (Minkowski space).

The embedding of the ordered field **Q** in the complete ordered field **R** suggests a method for embedding any metric space in a complete metric space as indicated in the following.

DEFINITION 5.2 A sequence $(P_n)$ in a metric space $\langle \mathbf{S}, \mathbf{D} \rangle$ is called *fundamental* in **S** if, for each $\varepsilon > 0$ in **R**, there is some $n_\varepsilon \in \mathbf{N}$ such that

$$\mathbf{D}(P_n, P_m) < \varepsilon \text{ in } \mathbf{R} \text{ for all } m, n \geqq n_\varepsilon \text{ in } \mathbf{N}.$$

DEFINITION 5.3 A sequence $(P_n)$ in a metric space $\langle \mathbf{S}, \mathbf{D} \rangle$ is *convergent* in $\mathbf{S}$ and has $P \in \mathbf{S}$ as a *limit* if for each $\varepsilon > 0$ in $\mathbf{R}$ there is some $n_\varepsilon \in \mathbf{N}$ such that

$$\mathbf{D}(P_n, P) < \varepsilon \text{ in } \mathbf{R} \text{ for all } m \geqq \text{ in } \mathbf{N}.$$

We write "$L(P_n) = P$ in $\mathbf{S}$" for "$(P_n)$ has $P$ as a limit."

THEOREM 5.1

(a) *Every sequence convergent in* $\mathbf{S}$ *is fundamental in* $\mathbf{S}$.

(b) *A convergent sequence in* $\mathbf{S}$ *has exactly one limit in* $\mathbf{S}$.

DEFINITION 5.4 A metric space $\mathbf{S}$ is *complete* if every sequence fundamental in $\mathbf{S}$ is convergent in $\mathbf{S}$.

THEOREM 5.2 *The metric space* $\mathbf{Q}^{(n)}$ *is not complete for any n. The Euclidean metric space* $\mathbf{R}^{(n)}$ *is complete for each n.*

We write $F_{\mathbf{S}}$ for the set of all sequences $(P_n)$ which are fundamental in the metric space $\mathbf{S}$.

THEOREM 5.3 *The set T of all* $((P_n), (Q_n)) \in F_{\mathbf{S}} \times F_{\mathbf{S}}$ *such that* $L(\mathbf{D}(P_n, Q_n)) = 0 \in \mathbf{R}$ *is an equivalence relation in* $F_{\mathbf{S}}$.

DEFINITION 5.5

(a) We call $(P_n)$ and $(Q_n)$ *equivalent* if $((P_n), (Q_n)) \in T$ and write "$(P_n) \sim (Q_n)$."

(b) $C_{(P_n)} = \{(Q_n) | (Q_n) \sim (P_n)\}$

(c) $\mathbf{S}^\star = \{P^\star | P^\star = C_{(P_n)} \text{ for some } (P_n) \in F_{\mathbf{S}}\}$

(d) If $P^\star = C_{(P_n)}, Q^\star = C_{(Q_n)}$ then $\mathbf{D}^\star(P^\star, Q^\star)$ $= L(\mathbf{D}(P_n, Q_n)) \in \mathbf{R}$.

THEOREM 5.4 *If* $(P_n) \sim (P'_n)$ *and* $(Q_n) \sim (Q'_n)$, *then there is exactly one* $\xi \in \mathbf{R}$ *such that*

$$L(\mathbf{D}(P_n, Q_n)) = \xi = L(\mathbf{D}(P'_n, Q'_n)).$$

PROOF: From the symmetry and triangle properties of the distance function $\mathbf{D}$ for the metric space $\mathbf{S}$ it follows that

(1) $|\mathbf{D}(P_n, Q_n) - \mathbf{D}(P_m, Q_m)| \leqq \mathbf{D}(P_n, P_m) + \mathbf{D}(Q_n, Q_m)$ in $\mathbf{R}$ for all $m, n \in \mathbf{N}$.

Since $(P_n)$, $(Q_n)$ are fundamental sequences in $\mathbf{S}$, it follows that $(\mathbf{D}(P_n, Q_n))$ is a fundamental sequence in $\mathbf{R}$. Since $\mathbf{R}$ is complete, there is some $\xi$ in $\mathbf{R}$ such that

(2) $L(\mathbf{D}(P_n, Q_n)) = \xi$ in $\mathbf{R}$.

Similarly,

(3) $L(\mathbf{D}(P'_n, Q'_n)) = \xi'$ in $\mathbf{R}$.

We have also, as in (1),

(4) $|\mathbf{D}(P_n, Q_n) - \mathbf{D}(P'_n, Q'_n)| \leqq \mathbf{D}(P_n, P'_n) + \mathbf{D}(Q_n, Q'_n)$ in $\mathbf{R}$ for all $n \in \mathbf{N}$.

By hypothesis, $L(\mathbf{D}(P_n, P'_n)) = 0 = L(\mathbf{D}(Q_n, Q'_n))$ in $\mathbf{R}$ and so, by (4), Hence

$$\xi - \xi' = L(\mathbf{D}(P_n, Q_n)) - L(\mathbf{D}(P'_n Q'_n))$$
$$= L(\mathbf{D}(P_n, Q_n) - \mathbf{D}(P'_n, Q'_n)) = 0,$$

and $\xi = \xi'$.

This theorem enables one to prove that $\mathbf{D}^\star$, defined in the next theorem, is a mapping which serves as a distance function for $\mathbf{S}^\star$.

THEOREM 5.5 *If* $P^\star = C_{(P_n)}, Q^\star = C_{(Q_n)} \in \mathbf{S}^\star$ *and* $\mathbf{D}^\star(P^\star, Q^\star) = L(\mathbf{D}(P_n, Q_n))$ *in* $\mathbf{R}$, *then*

(a) $\mathbf{D}^\star = \{((P^\star, Q^\star), \mathbf{D}^\star(P^\star, Q^\star)) \mid P^\star, Q^\star \in \mathbf{S}^\star\}$ *is a mapping of* $\mathbf{S}^\star \times \mathbf{S}^\star$ *into* $\mathbf{R}$.

(b) *The system* $\langle \mathbf{S}^\star, \mathbf{D}^\star \rangle$ *is a metric space.*

The metric space $\langle \mathbf{S}, \mathbf{D} \rangle$ may be embedded in the metric space $\langle \mathbf{S}^\star, \mathbf{D}^\star \rangle$ in the sense of the next theorem.

THEOREM 5.6 *The subset* $F$ *of* $\mathbf{S} \times \mathbf{S}^\star$ *defined by*

$$F = \{(P, C_{(P)}) \mid P \in \mathbf{S}\}$$

*is a mapping of* $\mathbf{S}$ *into* $\mathbf{S}^\star$ *such that*

$$\mathbf{D}(P, Q) = \mathbf{D}^\star(F(P), F(Q))$$

*for all* $P, Q \in \mathbf{S}$.

DEFINITION 5.6 If $\langle \mathbf{S}, \mathbf{D} \rangle$, $\langle \mathbf{S}', \mathbf{D}' \rangle$ are metric spaces, and $F$ is a 1–1 mapping of $\mathbf{S}$ into $\mathbf{S}'$ such that

$$\mathbf{D}(P, Q) = \mathbf{D}'(F(P), F(Q)),$$

then $F$ is called an *isometry* of $\mathbf{S}$ into $\mathbf{S}'$.

Thus the mapping $E_{\mathbf{Q}}^{\mathbf{R}}$ of Theorem 4.9 is an isometry of $\mathbf{Q}$ into $\mathbf{R}$.

THEOREM 5.7 $\langle \mathbf{S}^\star, \mathbf{D}^\star \rangle$ *is a complete metric space.*

The proof is analogous to that of Theorem 4.2, where $\mathbf{D}^\star$ and $\mathbf{D}$ replace "absolute value" in $\mathbf{R}$ and $\mathbf{Q}$, respectively. One establishes first the analog of

Theorem 4.1, proves "density" of **S** in $\mathbf{S}^{\star}$ in the sense of Corollary 1 of Theorem 4.10, and then proceeds to prove the completeness of $\mathbf{S}^{\star}$ as in Theorem 4.2

COROLLARY *Every metric space is isometric to a subspace of a complete metric space.*

# CHAPTER 5

# EQUIVALENT CHARACTERIZATIONS OF *R*

**1. DEFINITIONS.** By Theorem 4, 4.2, the ordered field **R** of real numbers is complete in the sense that every real fundamental sequence has a limit in **R**. By Theorem 4, 4.1, Corollary 3, **R** is an Archimedean ordered field. The completeness of **R** is one of several properties of the real number field which play a fundamental role in the theory of real-valued functions, and which have led to important generalizations in mathematics. In this chapter, we single out six properties in addition to completeness, and prove that an ordered field has any one of these properties if and only if it is Archimedean and complete. The ordered field **R**, being Archimedean as well as complete, has all of the properties.

We first introduce some terminology. In the following $\langle A, +, \cdot, < \rangle$ will be an ordered field.

DEFINITION 1.1

(1) A subset $X$ of $A$ is *bounded above* if there is an element $a \in A$ such that

$$x \leqq a \text{ for all } x \in X.$$

The element $a$ is called an *upper bound* for $X$.

(2) A subset $X$ of $A$ is *bounded below* if there is an element $a \in A$ such that

$$a \leqq x \text{ for all } x \in X.$$

The element $a$ is called a *lower bound* for $X$.

(3) A subset $X$ of $A$ is *bounded* if it has both an upper bound and a lower bound.

*Note:* $X$ is bounded if and only if there are elements $a_1, a_2 \in A$ such that

$$a_1 \leqq x \leqq a_2 \text{ for all } x \in X.$$

DEFINITION 1.2 An element $a$ of $A$ is called a *least upper bound* (*supremum*) of $X \in A$ if

(1) $a$ is an upper bound for $X$,
(2) $a \leqq u$ for all upper bounds $u$ of $X$.

An element $a$ of $A$ is called a *greatest lower bound* (*infimum*) of $X \subset A$ if

(1) $a$ is a lower bound for $X$,
(2) $v \leqq a$ for all lower bounds $v$ of $X$.

Note: A subset $X$ of $A$ has at most one least upper bound and at most one greatest lower bound.

*Notation:* We write "sup $X$" for the least upper bound of $X$ and "inf $X$" for the greatest lower bound of $X$. If sup $X \in X$ we write "max $X$" for "sup $X$." If inf $X \in X$ we write "min $X$" for "inf $X$."

DEFINITION 1.3 A subset $X$ of $A$ is called an *interval* in $A$ if, for some $a, b \in A$, one of the following holds:

(1) $X = \{x \mid a \leqq x \leqq b\}$
(2) $X = \{x \mid a < x < b\}$
(3) $X = \{x \mid a \leqq x < b\}$
(4) $X = \{x \mid a < x \leqq b\}$

The elements $a, b$ are called *endpoints* of the interval.
The element $b - a$ is *called* the *length* of the interval.
In the first case, we call $X$ a *closed interval* and write $X = [a, b]$.
In the second case, we call $X$ an *open interval* and write $X = (a, b)$.
In the third and fourth cases, $X$ is neither open nor closed. We write $X = [a,b)$ in the third case, and $X = (a, b]$ in the fourth case.
We note that if $b \leqq a$, then $(a, b)$, $[a, b)$, and $(a, b]$ are all empty, if $b < a$, then $[a, b]$ is empty, and if $b = a$, then $[a, b] = \{a\}$.

DEFINITION 1.4 An element $p \in A$ is called an *accumulation point* of $X \subset A$ if $X \cap [(a, b) - \{p\}] \neq \phi$ for all open intervals $(a, b)$ containing $p$.

DEFINITION 1.5 A subset $X$ of $A$ is *closed* if every accumulation point of $X$ is an element of $X$.

DEFINITION 1.6 A set $B$ of subsets $K$ of $A$ *covers* $X \subset A$ if for each $x \in X$ there is some $K \in B$ such that $x \in K$.

The following exercises illustrate the concepts introduced in Definitions 1.1 through 1.6.

*Exercise 1.1* If $X \subset A$ is bounded above, and $Y = \{x \mid -x \in X\}$, then

(1) $Y$ is bounded below,

(2) $a$ is an upper bound for $X$ if and only if $-a$ is a lower bound for $Y$.

*Exercise 1.2* A subset $X$ of $A$ has at most one least upper bound, and at most one greatest lower bound.

*Exercise 1.3* $a = \max X$ if and only if $x \leqq a \in X$ for all $x \in X$.
$a = \min X$ if and only if $x \geqq a \in X$ for all $x \in X$.

• *Exercise 1.4* If $a < b$, then any interval with endpoints $a$ and $b$ is an infinite set.

*Exercise 1.5* The element $p \in A$ is an accumulation point of $X \subset A$ if and only if $X \cap (a, b)$ is an infinite set for all open intervals $(a, b)$ containing $p$.

*Exercise 1.6* If $a < b$ in $A$, then $[a, b]$ is the set of all accumulation points of any interval with end points $a$ and $b$.

*Exercise 1.7* The subset $\{1 - 1/n \mid n = 1, 2, \ldots\}$ of $\mathbf{R}$ has 1 as its unique accumulation point. Does this statement generalize to arbitrary ordered fields?

*Exercise 1.8* If $X = Y \cup Z \subset \mathbf{R}$, where $Y = \{1/n \mid n = 1, 2, \ldots,\}$ and $Z = (1, 2)$, then $p$ is an accumulation point of $X$ if and only if $p = 0$ or $p \in [1, 2]$.

• *Exercise 1.9* For $a \in A$, $a \neq 0_A$, the set $X = \{na \mid n \in N\}$ is an infinite set which has no accumulation point.

*Exercise 1.10* The subset $X = \{x \mid x = 0 \text{ or } x = 1/n \text{ for } n \in \mathbf{N}\}$ of $\mathbf{R}$ is closed. Does this statement generalize to arbitrary ordered fields?

• *Exercise 1.11*

(a) If a subset $X$ of $A$ has no accumulation point in $A$, then $X$ is closed.

(b) No finite subset of $A$ has an accumulation point.

(c) If $Y \subset X \subset A$ and $x$ is an accumulation point of $X$ but not of $Y$, then $x$ is an accumulation point of $X - Y$.

(d) If $Y \subset X \subset A$ and $x$ is an accumulation point of $Y$, then $x$ is an accumulation point of $X$.

(e) The set $X$ of Exercise 1.9 is closed.

• *Exercise 1.12* If $a < b$, then the closed interval $[a, b]$ is a closed set, and the intervals $[a, b)$, $(a, b]$, and $(a, b)$ are not closed sets.

*Exercise 1.13* If for $a \in A$, $X_a = \{a\}$, then $\{X_a \mid a \in A\}$ covers $A$.

*Exercise 1.14* Let $X = (0, 1] \subset \mathbf{R}$ and let $K = \{(1/n, 2) \mid n \in \mathbf{N}\}$. Prove:

(1) $K$ covers $X$.

(2) No finite subset of $K$ covers $X$.

*Exercise 1.15* Let $X = [0, 1] \subset \mathbf{R}$ and let $K = \{(1/n, 2) \mid n \in \mathbf{N}\} \cup \{(1/2, 3/2)\}$.

(1) Prove that $K$ covers $X$.

(2) Find a finite subset of $K$ which covers $X$.

*Exercise 1.16*

(1) For every $e > 0$ in $\mathbf{Q}$, the set $B = \{(r - e, r + e) \mid r \in \mathbf{Q}\}$ covers $\mathbf{R}$.

(2) If $A_1$ and $A_2$ are ordered fields such that $A_1$ is dense in $A_2$, then for any $e > 0$ in $A_1$ the set $B = \{(r - e, r + e) \mid r \in A_1\}$ covers $A_2$.

*Exercise 1.17*

(a) Prove that, in $\mathbf{Q}$, the set $X = \{x \in \mathbf{Q} \mid x < 0 \text{ or } x^2 < 2\}$ is closed.

(b) Prove that, in $\mathbf{R}$, the set $\Xi = \{\xi \in \mathbf{R} \mid \xi < 0 \text{ or } \xi^2 < 2\}$ is not closed.

**2. EQUIVALENT PROPERTIES OF ORDERED FIELDS.** We now list six statements referring to an ordered field $\langle A, +, \cdot, < \rangle$ which we prove to be equivalent, in the sense that for a given ordered field, all of the statements are true if any one of them is true.

*Statement I*

(a) $A$ is Archimedean, and

(b) every fundamental sequence in $A$ has a limit in $A$.

*Statement II* Every non-empty subset of $A$ which is bounded above has a least upper bound in $A$.

*Statement III* There are no gaps in $A$.

*Statement IV* Every non-empty subset of $A$ which is bounded below has a greatest lower bound in $A$.

*Statement V* If $X$ is a bounded, closed subset of $A$ and $T$ is a set of open intervals which covers $X$, then $T$ has a finite subset $S$ which covers $X$.

*Statement VI* Every bounded infinite subset of $A$ has an accumulation point in $A$.

*Statement VII*

(a) $A$ is Archimedean, and

(b) if, for each $n \in \mathbf{N}$, $X_n$ is a closed interval in $A$, and $X_{n+1} \subset X_n$, then $\bigcap_{n \in \mathbf{N}} X_n \neq \phi$.

THEOREM 2.1 *In an ordered field $\langle A, +, \cdot, < \rangle$, Statements I through VII are equivalent.*

We prove the equivalence by establishing the following cycle of implication:

I IMPLIES II: If $A$ is Archimedean and every fundamental sequence in $A$ has a limit in $A$, then every non-empty subset of $A$ which is bounded above has a least upper bound in $A$.

PROOF: Suppose $X$ is a non-empty subset of $A$, $b$ is an upper bound for $X$, and $\bar{x} \in X$. Since $A$ is Archimedean, there is, for each $n \in \mathbf{N}$, some $\bar{m} \in \mathbf{N}$ such that $\bar{x} + \bar{m}/n \geqq b$ in $A$. Thus, $\bar{x} + \bar{m}/n$ is an upper bound for $X$. Hence, for each $n \in \mathbf{N}$, the set

$$B_n = \{ m \mid \bar{x} + m/n \text{ is an upper bound for } X \}$$

is a non-empty subset of $\mathbf{N}$, and (Theorem 1, 4.5) contains a smallest natural number, $m_n$. Then, for each $n \in \mathbf{N}$,

$$\text{(1)} \quad y_n = \bar{x} + \frac{m_n}{n} \text{ is an upper bound for } X$$

and

$$\text{(2)} \quad x_n = y_n - \frac{1}{n} = \bar{x} + \frac{m_n - 1}{n} < x \text{ in } A \text{ for some } x \in X.$$

Hence

$$x_m < y_n,$$

$$x_m - x_n < y_n - \left( y_n - \frac{1}{n} \right) = \frac{1}{n}.$$

and

$$|x_m - x_n| = \max \{x_m - x_n, x_n - x_m\}$$

$$\leqq \max \left\{ \frac{1}{n}, \frac{1}{m} \right\} \text{ in } A \text{ for all } m, n \in \mathbf{N}.$$

But then, since $L(1/n) = 0$ in the Archimedean ordered field $A$, $(x_n)$ is a fundamental sequence in $A$. By hypothesis, $(x_n)$ has a limit $a$ in $A$.

Now $a = \sup X$. For, $a$ is an upper bound for $X$. Otherwise, $a < x$ for some $x \in X$. Since $L(x_n) = a$ and $L(1/n) = 0$, there is some $n \in \mathbf{N}$ such that

$$x_n - a \leqq |x_n - a| < \frac{x - a}{2} \text{ and } \frac{1}{n} < \frac{x - a}{2} \text{ in } A.$$

Then, by (2),

$$y_n = x_n + \frac{1}{n} < a + \frac{x - a}{2} + \frac{x - a}{2} = x \text{ in } A.$$

This is impossible by (1), since $x \in X$. Furthermore, if $c$ is any upper bound for $X$, then $a \leqq c$ in $A$. Otherwise, $a - c > 0$ in $A$. Then, for some $n \in \mathbf{N}$,

$$a - x_n \leqq |a - x_n| < a - c \text{ in } A.$$

Hence, by (2), $c < x_n \leqq x$ in $A$ for some $x \in X$. This is impossible, since $c$ is an upper bound for $X$.

II IMPLIES III: If every non-empty subset of $A$ which is bounded above has a least upper bound in $A$, then no cut $(X, Y)$ in $A$ is a gap.

PROOF: Suppose $(X, Y)$ is a cut in $A$. Then $X$ is a non-empty subset of $A$ and every element of the non-empty set $Y$ is an upper bound for $X$ (Definition 1.2). By the hypothesis, there is some $a = \sup X$ in $A$.

Now, $a = \max X$ or $a = \min Y$. For, since $(X, Y)$ is a cut in $A$, $a \in X$ or $a \in Y$. If $a \in X$, then $\sup X = a = \max X$. If $a \in Y$, then, since every element of $Y$ is an upper bound for $X$, $\sup X = a = \min Y$.

III IMPLIES IV: If no cut $(X, Y)$ in $A$ is a gap, then every non-empty subset of $A$ which is bounded below has a greatest lower bound.

PROOF: Suppose $B$ is a non-empty subset of $A$ which is bounded below. Let

(1) $X = \{x \in A \mid x \leqq b \text{ for all } b \in B\}$.
(2) $Y = A - X$.

Then $(X, Y)$ is a cut in $A$. For, $X \neq \phi$ since $X$ is the set of all lower bounds for $B$, and $B$ is bounded below; $Y \neq \phi$ since $b + 1 \in Y$ for all $b$ in the non-empty set $B$; $X \cup Y = A$ and $X \cap Y = \phi$ by (2); and if $x \in X$, $y \in Y$, then $x < y$, since otherwise $y \leqq x \leqq b$ for all $b \in B$, and $y \in X$.

Suppose $X$ has no greatest element. Then $B \cap X = \phi$, for: if $b \in B \cap X$. then $b = \max X = \inf B$. Hence $B \subset Y$. Since $(X, Y)$ is a cut and there are no gaps in $A$, $Y$ has a least element, $y_0$. But, from $B \subset Y$, we have $y_0 \leqslant b$ for all

$b \in B$, hence $y_0 \in X$. This is impossible since $X \cap Y = \phi$. It follows that there is some $x_0 \in A$ such that $x_0 = \max X = \inf B$.

IV IMPLIES V: If every non-empty subset of $A$ which is bounded below has a greatest lower bound, then every set of open intervals which covers a bounded, closed subset $X$ of $A$ contains a finite subset which covers $X$.

PROOF: Let $X$ be a closed, bounded subset of $A$, and set $K$ be a set of open intervals which covers $X$. Since $X$ is bounded, there are elements $u, v \in A$ such that $X \subset [u, v]$. Since $X$ is closed, there is, for every $y \in [u, v]$, $y \notin X$, an open interval $J_y$ such that $y \in J_y$ and $X \cap J_y$ is empty. Let

$$H = \{J_y \mid y \in [u, v] - X, \text{and } J_y \cap X = \phi\}.$$

Then the set $M = K \cup H$ of open intervals $J \in K$ and $J_y \in H$ covers $[u, v]$. Now let

$$L = \{x \mid x \in [u, v] \text{ and } [x, v] \text{ is covered by a finite subset of } M\}.$$

Since $[v, v]$ is covered by some open interval $T_1 \in K$ if $v \in X$, or by $T_1 = J_v$ if $v \notin X$, it follows that $v \in L$ and $L$ is not empty. Since $u \leqq x$ for all $x \in L$, $L$ is bounded below. By Statement IV, $L$ has a greatest lower bound $x_0$.

We show that $[x_0, v]$ is covered by a finite subset of $M$, i.e., $x_0 \in L$. Since $x_0 \in [u, v]$, there is an open interval $T_0 \in M$ such that $x_0 \in T_0$. Let $T_0 = (a, b)$. Then $a < x_0 < b$. Since $x_0$ is the greatest lower bound of $L$, there is some $z_0 \in L$ such that $x_0 \leqq z_0 < b$. Since $z_0 \in L$, there is a finite set $\{T_1, \ldots, T_m\} \subset M$ which covers $[z_0, v]$. Hence the finite set $\{T_0, T_1, \ldots, T_m\} \subset M$ covers $[x_0, v]$.

We show next that $x_0 = u$. If $x_0 \neq u$, then $u < x_0$. Since $a < x_0$, $\max\{u, a\} < x_0$. Since the order in $A$ is dense (Theorem 3, 4.3), there is an element $z_1 \in A$ such that

$$u, a \leqq \max\{u, a\} < z_1 < x_0 \leqq \min\{b, v\} \leqslant v.$$

Hence, $z_1 \in [u, v]$, and $[z_1, v]$ is covered by $\{T_0, T_1, \ldots, T_m\}$. Thus, $z_1 \in L$, and $z_1 < x_0$. This is impossible, since $x_0$ is the greatest lower bound of $L$.

We have shown that $[u, v]$, and hence the subset $X$ of $[u, v]$, is covered by $\{T_0, T_1, \ldots, T_m\} \subset M = K \cup H$. Since no interval in $H$ contains a point of $X$, $K \cap \{T_0, T_1, \ldots, T_m\}$ is a finite subset of $K$ which covers $X$.

V IMPLIES VI: If every set of open intervals which covers a bounded, closed subset $X$ of $A$ contains a finite subset which covers $X$, then every bounded, infinite subset of $A$ has an accumulation point in $A$.

PROOF: Let $X$ be a bounded infinite subset of $A$, and suppose $X$ has no accumulation point. Then $X$ is closed. Suppose $x \in X$. Since $x$ is not an accumulation point of $X$, there is an open interval $J_x$ such that $X \cap J_x = \{x\}$. The set $K = \{J_x | x \in X\}$ covers $X$. Since $X$ is closed and bounded, there is a finite subset $\overline{K}$ of $K$ which covers $X$. If $\overline{K} = \{J_{x_1}, \ldots, J_{x_n} | n \in \mathbf{N}\}$ then for every $x \in X$, there is some $k \in I_n$ such that $x \in J_{x_k}$ and hence $x = x_k$. But then $X = \{x_1, \ldots, x_n\}$, a finite set, contrary to hypothesis.

VI IMPLIES VII: If every bounded, infinite subset of $A$ has an accumulation point in $A$, then (a) $A$ is Archimedean and (b) if $(J_n)$ is a sequence of closed intervals in $A$ such that $J_{n+1} \subset J_n$ for all $n \in \mathbf{N}$, then $\bigcap_{n \in \mathbf{N}} J_n \neq \phi$.

PROOF: (a) If $A$ is not Archimedean, there exist $a, b \in A$ such that $0 < a < b$ in $A$, and $a \leqq na < b$ for all $n \in \mathbf{N}$. Hence, the set $X = \{na | n \in \mathbf{N}\}$ is bounded and infinite, and has no accumulation point (Exercise 1.9), contrary to the hypothesis. Thus $A$ is Archimedean.

(b) Suppose $J_n = [a_n, b_n]$ and $J_{n+1} \subset J_n$ for each $n \in \mathbf{N}$. Then $a_n \leqq a_{n+1} \leqq b_{n+1} \leqq b_n$ for each $n$. Let $X = \{a_n | n \in \mathbf{N}\}$. Since $a_1 \leqq a_n \leqq b_1$ for all $n \in \mathbf{N}$, $X$ is a bounded subset of $A$.

If for some $\bar{n} \in \mathbf{N}$, $a_m = a_{\bar{n}}$ for all $m \geqq \bar{n}$, then $a_n \leqq a_{\bar{n}} \leqq b_n$ for all $n$, and $a_{\bar{n}} \in \bigcap_{n \in \mathbf{N}} J_n$. Otherwise, for each $n \in \mathbf{N}$ there is some $m \in \mathbf{N}$ such that $a_m > a_n$. Hence the set $X$ has no greatest element and is therefore an infinite subset of $A$. By the hypothesis, the bounded, infinite set $X$ has an accumulation point $x \in A$.

If $a_n > x$ for some $n$, then $a_m \geqq a_n > x$ for all $m \geqq n$. Thus, if $0 < \varepsilon < a_n - x$, the interval $(x - \varepsilon, x + \varepsilon)$ contains only a finite number of points of $X$. Hence $a_n \leqq x$ for all $n$. If $b_n < x$ for some $n$, then $a_m \leqq b_n < x$ for all $m$. Thus, if $0 < \varepsilon < x - b_n$, the interval $(x - \varepsilon, x + \varepsilon)$ contains no points of $X$. Hence $x \leqq b_n$ for each $n$. But then $a_n \leqq x \leqq b_n$ for each $n$, and $x \in \bigcap_{n \in N} J_n$.

VII IMPLIES I: If (a) $A$ is Archimedean and (b) if $(J_n)$ is a sequence of closed intervals in $A$ such that $J_{n+1} \subset J_n$ for all $n \in \mathbf{N}$, then $\bigcap_{n \in \mathbf{N}} J_n \neq \phi$, then (a′) $A$ is Archimedean and (b′) every fundamental sequence in $A$ has a limit in $A$.

PROOF: (a′) holds by (a) of the hypothesis. Suppose $(x_n)$ is a fundamental sequence in $A$. Then for each $k \in \mathbf{N}$ there is some $n_k \in \mathbf{N}$ such that

(1) $x_{n_k} - 1/k < x_n < x_{n_k} + 1/k$ in $A$ for all $n \geqq n_k$ in $\mathbf{N}$.

For each $m \in \mathbf{N}$ there are

(2) $p_m = \max\{n_k | k \leqq m \text{ in } \mathbf{N}\} \in \mathbf{N}$,

(3) $a_m = \max\{x_{n_k} - 1/k \mid k \leqq m \text{ in } \mathbf{N}\} \in A$,
(4) $b_m = \min\{x_{n_k} + 1/k \mid k \leqq m \text{ in } \mathbf{N}\} \in A$.

Then, by (2), (3), and (4)

(5) $a_m \leqq a_{m+1} < x_n < b_{m+1} \leqq b_m$ in $A$
for all $m \in \mathbf{N}$ and all $n \geqq p_{m+1}$ in $\mathbf{N}$.

Let $J_m = [a_m, b_m]$ for each $m \in \mathbf{N}$. Then, by (5), $J_{m+1} \subset J_m$ for all $m \in \mathbf{N}$. Hence, by hypothesis (b), there is some $a \in A$ such that

(6) $a \in \bigcap_{m \in \mathbf{N}} J_m$

Now $L(x_n) = a$. For, by (1), (3), (4), and (6),

(7) $x_{n_m} - 1/m \leqq a_m \leqq a \leqq b_m \leqq x_{n_m} + 1/m$ in $A$
for all $m \in \mathbf{N}$.

By (1),

(8) $x_{n_m} - 1/m < x_n < x_{n_m} + 1/m$ in $A$ for all $n \geqq n_m$ in $\mathbf{N}$.

Hence, by (7) and (8),

(9) $|x_n - a| = \max\{x_n - a, a - x_n\} \leqq 2/m$ in $A$
for all $m \in \mathbf{N}$ and all $n \geqq n_m$ in $\mathbf{N}$.

Let $e > 0$ in $A$. By (a), there is some $m \in \mathbf{N}$ such that

$$me > 2 \text{ in } A$$

Hence, by (9),

$$|x_n - a| < e \text{ in } A \text{ for all } n \geqq n_m \text{ in } \mathbf{N},$$

and $L(x_n) = a$.

We note that the Archimedean property (Statement I (a)) is not a consequence of the completeness of an ordered field, i.e., Statements I (b) and VII (b) are not equivalent (see Exercise 2.2). Examples also exist showing that the Archimedean property (Statement VII (a)) is not a consequence of the "nested intervals" property (Statement VII (b)) (cf. [8]). Thus, the hypothesis that the ordered field is Archimedean is actually needed in Statements I and VII.

*Exercise 2.1* Find a counter example in $\mathbf{Q}$ to each of Statements I through VII, based on the non-existence in $\mathbf{Q}$ of a square root of 2.

*Exercise 2.2* Let $K$ be the set of all Laurent series $\sum_{-\infty}^{\infty} \alpha_j x^j$ where $\alpha_j \in \mathbf{R}$ for each $j$, and for some integer $h$, $\alpha_j = 0$ if $j < h$. If addition, multiplication and order are defined as in Exercise 3, 4.5, then

(a) $K$ forms a non-Archimedean ordered field which is complete;

(b) A set of "nested" closed intervals–i.e., intervals satisfying the hypotheses of Statement VII (b)–may have an empty intersection. Hint: consider the set of intervals $J_n$ given by $\left[\sum_{j=1}^{\infty} nx^j, \frac{1}{n}\right]$.

3. **CATEGORICITY.** We now show that any one of Statements I through VII characterizes **R** among all ordered fields, or, that the statement "$A$ is an ordered field" together with any one of Statements I through VII forms a categorical set of axioms for the ordered field of real numbers. Because of the equivalence of the seven statements, it is sufficient to prove that any one of the statements characterizes **R** among ordered fields.

We first prove:

THEOREM 3.1 *Any Archimedean ordered field can be isomorphically embedded in the ordered field* **R** *of real numbers.*

PROOF: Let $A$ be an Archimedean ordered field. Let $\mathbf{Q}_A$ be the set of all rational elements of $A$, and let $\tilde{x}$ be the rational element of $A$ which corresponds to $x \in \mathbf{Q}$ under the embedding of Theorem 3, 4.5.

If $F\colon A \to \mathbf{R}$ is defined by $F(L(x_n)) = L(x_n)$ for each $(x_n) \in F_{\mathbf{Q}}$, then $F$ is an injection. For, if $a \in A$, then, by Exercise 4, 4.3, $a = L(\tilde{x}_n)$ for some sequence $(\tilde{x}_n)$ in $\mathbf{Q}_A$. Since $(\tilde{x}_n)$ is a fundamental sequence in $A$ (Theorem 3, 5.6), $(\tilde{x}_n)$ is a fundamental sequence in $\mathbf{Q}_A$, and $(x_n)$ is a fundamental sequence in $\mathbf{Q}$ and in **R** (Exercises 3, 5.14, 3, 4.2). Since **R** is complete, $L(x_n) \in \mathbf{R}$, and $(a, L(x_n)) \in F$. For any $(x_n), (y_n) \in F_{\mathbf{Q}}$,

(1) $L(x_n) = L(y_n)$ in **R**

if and only if (Exercise 3, 4.2)

(2) $L(x_n - y_n) = 0$ in $\mathbf{Q}$.

But (2) holds if and only if (Exercise 3, 5.14)

(3) $L(\tilde{x}_n - \tilde{y}_n) = 0$ in $\mathbf{Q}_A$,

and (3) holds if and only if (Exercise 3, 4.2)

(4) $L(\tilde{x}_n) = L(\tilde{y}_n)$ in $A$.

Thus, (1) and (4) are equivalent, and $F$ is an injection from $A$ to **R**. By Theorem 3, 5.12, $F(a +_A b) = F(a) +_{\mathbf{R}} F(b)$ and $F(a \cdot_A b) = F(a) \cdot_{\mathbf{R}} F(b)$ for all $a, b \in A$. Thus, $F$ preserves addition and multiplication. If $a = L(\tilde{x}_n)$ is a positive element in $A$, then $(\tilde{x}_n)$ is a positive sequence in $A$

(Exercise 3, 5.14). By Exercises 3, 5.13 and 3, 5.14, $(x_n)$ is a positive sequence in **R**, and $L(x_n)$ is a positive real number. By Exercise 3, 3.1, $F$ preserves order.

THEOREM 3.2 *Any complete Archimedean ordered field is isomorphic to the ordered field of real numbers.*

PROOF: If $A$ is complete, the function $F$ of Theorem 3.1 is a bijection. For, if $x = L(x_n)$ in **R**, then $(\tilde{x}_n)$ is a fundamental sequence in $A$, and, since $A$ is complete, $L(\tilde{x}_n) = a \in A$. But then $x = F(a)$. Hence, $F$ is an isomorphism of the ordered field $A$ onto the ordered field **R**.

COROLLARY *Any ordered field satisfying one of Statements* I *through* VII *is isomorphic to* **R**.

**4. UNCOUNTABILITY OF R.** Finally, we prove that the ordered field **R** of real numbers is more numerous than **N**, **Z**, or **Q**.

THEOREM 4.1 **R** *is not countable.*

PROOF: Suppose **R** is countable. Then there is a function $F$: $\mathbf{N} \to \mathbf{R}$ such that

(1) $\mathbf{R} = \{\xi_n = F(n) \mid n \in \mathbf{N}\}$.

Let $A$ be the set of all closed intervals $[\eta, \zeta]$ such that $\eta < \zeta$ in **R**. For each $[\eta, \zeta] \in A$ and $n \in \mathbf{N}$, let

$$\eta_1 = \eta,\ \ \eta_2 = \eta + (\zeta - \eta)/3,\ \ \eta_3 = \eta + 2(\zeta - \eta)/3,\ \ \eta_4 = \zeta,$$

and let $\eta_p$ be the smallest of $\eta_1, \eta_2, \eta_3$ such that $\xi_{n+1} \notin [\eta_p, \eta_{p+1}]$. Note that $[\eta_p, \eta_{p+1}]$ is the first third of $[\eta, \zeta]$ from the left which excludes $\xi_{n+1}$.

For each $n \in \mathbf{N}$, define $G_n$: $A \to A$ by: $G_n([\eta, \zeta]) = [\eta_p, \eta_{p+1}]$ for each $[\eta, \zeta] \in A$. By the Generalized Recursion Theorem (Exercise 1, 2.4), there is a sequence $(J_n)$ of closed intervals such that

(2) $J_1 = [\xi_1 + 1, \xi_1 + 2]$ and $J_{n+1} = G_n(J_n)$ for each $n \in \mathbf{N}$.

Then

(3) $\xi_n \notin J_n$ and $J_{n+1} \subset J_n$ for each $n \in \mathbf{N}$.

By (3) and the nested intervals property of **R**,

(4) $\xi \in \bigcap_{n \in \mathbf{N}} J_n$ for some $\xi \in \mathbf{R}$.

By (1) and (3), $\xi = \xi_k \notin J_k$ for some $k \in \mathbf{N}$. Hence $\xi \notin \bigcap_{n \in \mathbf{N}} J_n$, contrary to (4). It follows that **R** is not countable.

*Exercise 4.1* If $x$ is a real number in the interval $(0, 1]$, prove that $x$ can be represented in exactly one way by a "non-terminating decimal $.a_1a_2a_3 \ldots$," i.e., prove that $x$ is the limit of a unique sequence $(u_n)$ of the form

$$u_1 = \frac{a_1}{10}$$

$$u_n = u_{n-1} + \frac{a_n}{10^n}, n > 1,$$

where $0 \leqq a_n \leqq 9$ and $u_n < x$ for each $n$.

*Exercise 4.2* Using the representation in Exercise 4.1, prove that $(0, 1]$, and hence **R** (see Ex. 1.21), is uncountable.

Hint: suppose $(0, 1]$ is countable, and list its elements

$$\begin{array}{l} x_1\colon .a_{11}a_{12}a_{13} \ldots \\ x_2\colon .a_{21}a_{22}a_{23} \ldots \\ x_3\colon .a_{31}a_{32}a_{33} \ldots \\ \cdot \quad \cdot \quad \cdot \quad \cdot \\ \cdot \quad \cdot \quad \cdot \quad \cdot \\ \cdot \quad \cdot \quad \cdot \quad \cdot \end{array}$$

Construct a non-terminating decimal $.b_1b_2 \ldots$ such that $b_n \neq a_{nn}$ for each $n$, and obtain a contradiction.

# CHAPTER 6

# THE COMPLEX NUMBERS

## 1. THE FIELD OF COMPLEX NUMBERS.

The stages in our construction of the real number system may be viewed from the standpoint of the solution of polynomial equations. In the natural number system **N**, an equation of form

(1) $x + n = m$

has no root for $n \geqq m$. In the domain **Z** of integers, any equation of form (1) has a root, but an equation of form

(2) $ax = b, a \neq 0, a, b \in \mathbf{Z}$,

generally has no root in **Z**. (When it does, we say that $a$ is a divisor of $b$.) This situation is remedied in the field **Q** of rational numbers where all equations of form

(2′) $px = q, p \neq 0, p, q \in \mathbf{Q}$,

have roots. However, for example, an equation of form

(3) $x^n = z, n \in \mathbf{N}, z \in \mathbf{Q}$,

may have no root in **Q**. (We have discussed in detail the case $n = 2$, $z = 2$ (Theorem 3, 5.1 (1).)

This situation is partially remedied in the field **R** of real numbers where any equation of form

(3′) $x^n = \xi, \xi \in \mathbf{R}$,

has a root if $n$ is odd, but fails to have a root if $n$ is even and $\xi$ is negative, since even powers in ordered fields are positive. To remedy this situation we construct yet another extension. Our immediate goal shall be to embed **R** in a field in which the equation

(4) $x^2 = -1$

has a root. The resulting field will, in fact, far surpass this goal. For, the field of complex numbers which we are about to construct will have the property that *every* polynomial equation with coefficients in the field has a root* in the field. A field with this property is called *algebraically closed*. The theorem which asserts that the field of complex numbers is algebraically closed is traditionally called the Fundamental Theorem of Algebra, although its proof cannot be accomplished by algebraic means alone. This theorem was first proved by Gauss in his doctoral dissertation. (Gauss later gave at least six additional proofs of the theorem.)

DEFINITION 1.1 We denote by $\mathbf{C}$ the set $\mathbf{R} \times \mathbf{R}$ of all ordered pairs $(\xi, \eta)$ of real numbers, write $u, v, w, z, \ldots$ for the elements of $\mathbf{C}$, and call them *complex numbers*.

(Observe that the usual equivalence relation (pp. 42, 58, 79) in this case is simply equality of ordered pairs, so that each equivalence class consists of a single element.)

THEOREM 1.1 *If, for* $u = (\xi, \eta)$ *and* $v = (\sigma, \tau)$,

$$F(u, v) = (\xi + \sigma, \eta + \tau)$$

*and*

$$G(u, v) = (\xi\sigma - \eta\tau, \xi\tau + \sigma\eta),$$

*then*

$$F = \{((u, v), F(u, v)) \mid u, v \in \mathbf{C}\}$$

*and*

$$G = \{((u, v), G(u, v)) \mid u, v \in \mathbf{C}\}$$

*are binary operations on* $\mathbf{C}$.

PROOF: Since two ordered pairs of real numbers are equal if and only if they agree in each component, $F(u, v)$ and $G(u, v)$ are uniquely determined for each $(u, v) \in \mathbf{C} \times \mathbf{C}$. Hence $F$ and $G$ are mappings of $\mathbf{C} \times \mathbf{C}$ into $\mathbf{C}$, i.e., they are binary operations on $\mathbf{C}$.

DEFINITION 1.2 We call $F$ and $G$, respectively, *addition* and *multiplication* on $\mathbf{C}$, write $u +_{\mathbf{C}} v$ for $F(u, v)$ and $u \cdot_{\mathbf{C}} v$ for $G(u, v)$, and omit the subscript "$\mathbf{C}$" most of the time.

THEOREM 1.2 *The system* $\langle \mathbf{C}, +_{\mathbf{C}}, \cdot_{\mathbf{C}} \rangle$ *is a field.*

*In fact, all of its roots.

PROOF: We leave to the reader the verification that $\langle \mathbf{C}, +_{\mathbf{C}}, \cdot_{\mathbf{C}} \rangle$ is a commutative ring with additive identity $0_{\mathbf{C}} = (0, 0)$ and multiplicative identity $1_{\mathbf{C}} = (1, 0)$. To show that **C** is, in fact, a field, we observe that if $u = (\xi, \eta) \neq 0_{\mathbf{C}}$, then either $\xi \neq 0$ or $\eta \neq 0$. Hence $\xi^2 + \eta^2$ is a positive real number. But then

$$v = \left(\frac{\xi}{\xi^2 + \eta^2}, \frac{-\eta}{\xi^2 + \eta^2}\right)$$

is an element of **C**, and

$$uv = (\xi, \eta)\left(\frac{\xi}{\xi^2 + \eta^2}, \frac{-\eta}{\xi^2 + \eta^2}\right) = (1, 0) = 1_{\mathbf{C}}.$$

Thus, every non-zero element of **C** has a multiplicative inverse, and $\langle \mathbf{C}, +_{\mathbf{C}}, \cdot_{\mathbf{C}} \rangle$ is a field.

THEOREM 1.3 *The complex numbers* $i = (0, 1)$ *and* $-i = (0, -1)$ *are solutions of the equation* $x^2 = -1_{\mathbf{C}}$.

PROOF: $(\pm i)^2 = i^2 = (0, 1)(0, 1) = (-1, 0) = -1_{\mathbf{C}}$.

*Exercise 1.1* The equation $x^2 = -1$ has no solution other than $\pm i$ in **C**.

*Exercise 1.2*

(1) If $(\xi, \eta) < (\xi', \eta')$ whenever $\xi < \xi'$ or $\xi = \xi'$ and $\eta < \eta'$ in **R**, then $<$ is an order relation in **C**.

(2) Is $\langle \mathbf{C}, +, \cdot, < \rangle$ an ordered field?

(3) Is there any order relation, $<$, such that $\langle \mathbf{C}, +, \cdot, < \rangle$ is an ordered field?

## 2. EMBEDDING.

THEOREM 2.1 *If* $E: \mathbf{R} \to \mathbf{C}$ *is defined by* $E(\xi) = (\xi, 0)$ for each $\xi \in \mathbf{R}$, *then E is an isomorphic embedding of the field* $\langle \mathbf{R}, +_{\mathbf{R}}, \cdot_{\mathbf{R}} \rangle$ into the *field* $\langle \mathbf{C}, +_{\mathbf{C}}, \cdot_{\mathbf{C}} \rangle$.

PROOF: For every $\xi \in \mathbf{R}$, there is exactly one $u \in \mathbf{C}$ such that $u = (\xi, 0) = E(\xi)$. Thus, $E$ is a mapping of **R** into **C**. Since $(\xi, 0) = (\eta, 0)$ only if $\xi = \eta$, $E$ is a 1-1 mapping of **R** into **C**. Since $E(\xi +_{\mathbf{R}} \eta) = (\xi +_{\mathbf{R}} \eta, 0) = (\xi, 0) +_{\mathbf{C}} (\eta, 0)$, and

$$E(\xi \cdot_{\mathbf{R}} \eta) = (\xi, 0) \cdot_{\mathbf{C}} (\eta, 0) \text{ for all } \xi, \eta \in \mathbf{R},$$

$E$ is an isomorphic embedding of $\langle \mathbf{R}, +_{\mathbf{R}}, \cdot_{\mathbf{R}} \rangle$ into $\langle \mathbf{C}, +_{\mathbf{C}} \cdot_{\mathbf{C}} \rangle$.

In view of Theorem 2.1, we shall write "$\xi$" for $E(\xi) = (\xi, 0)$ for all $\xi \in \mathbf{R}$. This will permit us to express the elements of **C** in the conventional notation for complex numbers.

THEOREM 2.2 *If $z \in \mathbf{C}$, then z can be expressed in one and only one way as $z = \xi + \eta i$, where $\xi, \eta \in \mathbf{R}$ and $i = (0, 1)$.*

PROOF: Since $z \in \mathbf{C}$, $z = (\xi, \eta)$ for some $\xi, \eta \in \mathbf{R}$. Hence $z = (\xi, \eta) = (\xi, 0) + (0, \eta) = (\xi, 0) + (\eta, 0)(0, 1) = \xi + \eta i$.

If $z = \xi' + \eta' i$ for $\xi', \eta' \in \mathbf{R}$, then $\xi' + \eta' i = (\xi', 0) + (\eta', 0)(0, 1) = (\xi', 0) + (0, \eta') = (\xi', \eta') = z = (\xi, \eta)$. Hence, $\xi = \xi'$ and $\eta = \eta'$.

**3. C AS A VECTOR SPACE.** The last theorem exemplifies yet another aspect of **C**. Ordered pairs of real numbers, or, equivalently, complex numbers, are traditionally represented as points, or vectors, in the Cartesian plane. Addition of complex numbers, in this representation, corresponds to vector addition, and multiplication of complex numbers by real numbers corresponds to the multiplication of vectors by real scalars. The properties of vectors with respect to these two operations are formalized in the definition of a vector space.

DEFINITION 3.1 If $\langle K, +_K, \cdot \rangle$ is a field, $\langle V, +_V \rangle$ is a commutative group and $\circ$ is a binary operation on $K \times V$ into $V$, then *V is a vector space (linear space) over K* if

(1) $\alpha \circ (a +_V b) = \alpha \circ a +_V \alpha \circ b$
(2) $(\alpha +_K \beta) \circ a = \alpha \circ a +_V \beta \circ a$
(3) $(\alpha\beta) \circ a = \alpha \circ (\beta a)$
(4) $1_K \circ a = a$

for all $\alpha, \beta \in K$ and all $a, b \in V$.

The operation $\circ$ is called *scalar multiplication*; the addition in $V$ is called *vector addition.* (Strictly, the system

$$\langle\langle K, +_K, \cdot \rangle, \langle V, +_V \rangle, \circ \rangle$$

is the vector space!)

DEFINITION 3.2 If $V$ is a vector space over a field $K$, then the $n$-tuple $\langle a_1, \ldots, a_n \rangle$ of vectors $a_i$ of $V$ is called a *basis* for $V$ if every vector $v \in V$ has a unique representation.

$$v = \sum_{i=1}^{n} \alpha_i a_i$$

where $\alpha_i \in K$ for each $i = 1, \ldots, n$.

DEFINITION 3.3 If $V$ is a vector space over $K$, and $n \in \mathbf{N}$, then $V$ *has dimension n* if it has a basis of $n$ elements.

It is proved in any standard treatment of vector spaces that any two bases of a vector space have the same number of elements, so that the dimension of a finite dimensional vector space is uniquely determined (cf. [2]).

THEOREM 3.1 *$\mathbf{C}$ is a two-dimensional vector space over* $\mathbf{R}$.

PROOF: We leave to the reader the verification that $\mathbf{C}$ satisfies the conditions of Definition 3.3. That $\mathbf{C}$ is two-dimensional follows immediately from Theorem 2.2. For, according to Theorem 2.2, $1_{\mathbf{C}} = (1, 0)$ and $i = (0, 1)$ serve as a basis for the vector space $\mathbf{C}$ over $\mathbf{R}$.

DEFINITION 3.4 If the elements of a vector space $V$ over a field $K$ form a ring $\langle V, +, \cdot \rangle$ such that

$$\alpha(ab) = (\alpha a)b = a(\alpha b)$$

for all $\alpha \in K$ and all $a, b \in V$, then $V$ is called an *algebra* over $K$. If the vector space $V$ has dimension $n$ over $K$, then the algebra $V$ has dimension $n$ over $K$.

COROLLARY $\mathbf{C}$ is an algebra over $\mathbf{R}$, of dimension 2.

PROOF: See Exercise 3.4.

We state, but do not prove, the following remarkable theorem (Weierstrass-Frobenius Theorem), cf. [18]): Any field which is a finite dimensional algebra over $\mathbf{R}$ is isomorphic either to $\mathbf{C}$ or to $\mathbf{R}$.

*Exercise 3.1*

(1) If $K$ is any field, $K_{\mathbf{N}}$ the set of all sequences $(a_n)$ in $K$, and if addition and scalar multiplication are defined by

$$(a_n) + (b_n) = (a_n + b_n)$$

$$\alpha \circ (a_n) = (\alpha a_n)$$

then $K_{\mathbf{N}}$ is a vector space over $K$.

(2) If $K$ is an ordered field, and $V$ is the set
- (a) of all fundamental sequences in $K$,
- (b) of all convergent sequences in $K$,
- (c) of all sequences with limit 0 in $K$,

and the operations are defined as in (1), then, in each case, $V$ is a vector space over $K$.

*Exercise 3.2* If $K$ is any field, $n$ is a fixed natural number, and the operations are defined componentwise, then the set $K_n$ of all $n$-tuples over $K$ is an $n$-dimensional vector space over $K$.

*Exercise 3.3* Every field is a 1-dimensional vector space over itself.

*Exercise 3.4* Prove that **C** is an algebra of dimension 2 over **R**.

**4. C AS A METRIC SPACE.** Since **C** cannot be made into an ordered field, the usual definition of absolute value is not possible for complex numbers. However, we shall define a mapping of **C** into **R** which shares some of the properties of the absolute value function we defined for ordered fields.

DEFINITION 4.1 If $z = (\xi, \eta) \in \mathbf{C}$, then the non-negative real number $\rho(z)$ such that $\xi^2 + \eta^2 = \rho^2(z)$ is called the modulus (absolute value) of $z$.

THEOREM 4.1 *If, for all $z, w \in \mathbf{C}$, $\delta(z, w) = \rho(z - w)$, then the set*

$$\delta = \{((z, w), \delta(z, w)) \mid (z, w) \in \mathbf{C} \times \mathbf{C}\}$$

*is a metric on **C**, and $\langle \mathbf{C}, \delta \rangle$ is a metric space.*

PROOF: By Exercise 4.1, since $z - w$ is uniquely determined for each $(z, w) \in \mathbf{C} \times \mathbf{C}$, $\delta$ is a function from $\mathbf{C} \times \mathbf{C}$ to **R**. By Exercise 4.2, (1), $\delta(z, w) \geqq 0$ for all $(z, w) \in \mathbf{C} \times \mathbf{C}$, and $\delta(z, w) = 0$ if and only if $z = w$. By Exercise 4.2 (2),

$$\delta(z, w) = \rho(z - w) = \rho(w - z) = \delta(w, z) \text{ for all } z, w \in \mathbf{C}.$$

By Exercise 4.2, (3),

$$\delta(z, v) + \delta(v, w) = \rho(z - v) + \rho(v - w) \geqq \rho(z - w) = \delta(z, w).$$

Thus, by Definition 4, 5.1, $\delta$ is a metric on **C**, and $\langle \mathbf{C}, \delta \rangle$ is a metric space.

In Chapter 4, we gave definitions of fundamental sequences, convergence, and completeness for metric spaces.

THEOREM 4.2 *$\langle \mathbf{C}, \delta \rangle$ is a complete metric space, i.e., every fundamental sequence in **C** has a limit in **C**.*

PROOF: Let $(z_n)$ be a fundamental sequence in **C**. Then for each $\varepsilon > 0$ there is some $n_\varepsilon \in \mathbf{N}$ such that

(1) $\delta(z_m, z_n) < \varepsilon$

for all $m, n \geqq n_\varepsilon$ (Definition 4, 5.2). If $z_n = (\xi_n, \eta_n)$ for each $n$, then, by Exercise 4.4,

(2) $|\xi_m - \xi_n| \leqq \delta(z_m, z_n) < \varepsilon$

and

(3) $|\eta_m - \eta_n| \leqq \delta(z_m, z_n) < \varepsilon$

for all $m, n \geqq n_\varepsilon$.

Hence, the sequences $(\xi_n)$ and $(\eta_n)$ are fundamental in **R**. By the completeness of **R** (Theorem 4, 4.2), there are real numbers $\xi$, $\eta$ such that

$$L(\xi_n) = \xi \quad \text{and} \quad L(\eta_n) = \eta.$$

Hence, for every $\varepsilon > 0$, there are natural numbers $n'_\varepsilon$, $n''_\varepsilon$ such that

$$|\xi_n - \xi| < \frac{\varepsilon}{\sqrt{2}} \text{ for } n \geqq n'_\varepsilon$$

and

$$|\eta_n - \eta| < \frac{\varepsilon}{\sqrt{2}} \text{ for } n \geqq n''_\varepsilon.$$

But then for $n_\varepsilon = \max\{n'_\varepsilon, n''_\varepsilon\}$,

$$\delta(z, z_n) \leqq \sqrt{2} \max\{|\xi - \xi_n|, |\eta - \eta_n|\} < \varepsilon$$

for all $n \geqq n_\varepsilon$.

Hence, the sequence $(z_n)$ converges to $z \in$ **C**.

● *Exercise 4.1* The set $\rho = \{(z, \rho(z)) | z \in \mathbf{C}\}$ is a mapping of **C** into **R**.

● *Exercise 4.2*

(1) If $z \in \mathbf{C}$, then $\rho(z) = 0$ if and only if $z = 0$.
(2) $\rho(-z) = \rho(z)$ for all $z \in \mathbf{C}$.
(3) For $z, w \in \mathbf{C}$, $\rho(zw) = \rho(z)\rho(w)$.
(4) For $z, w \in \mathbf{C}$, $\rho(z + w) \leqq \rho(z) + \rho(w)$.
(5) For $z, w \in \mathbf{C}$, $\rho(z - w) \geqq |\rho(z) - \rho(w)| \geqq \rho(z) - \rho(w)$.

*Exercise 4.3* The isomorphic embedding $E$ of Theorem 2.1 is an isometry of (**R**, **D**) (Definition 4, 5.6) into $\langle \mathbf{C}, \boldsymbol{\delta} \rangle$.

*Exercise 4.4* If $z = (\xi, \eta)$ and $w = (\sigma, \tau)$ then

$$\max\{|\xi - \sigma| \; |\eta - \tau|\} \leqq \delta(z, w) \leqq \sqrt{2} \max\{|\xi - \sigma|, |\eta - \tau|\}.$$

# Bibliography

Completeness is not a property of the bibliography. Where a bracketed number appears in the text, it is a reference to an item in the following list.


1 H. Bachman, Transfinite Zahlen, *Ergebnisse der Math. und ihrer Grenzgebiete,* Neue Folge, Heft 1, Springer Verlag, 1955

2 G. Birkhoff and S. MacLane, *A Survey of Modern Algebra,* Macmillan, 1956

3 N. Bourbaki, *Eléments de Mathématique,* Première partie, Livre I, Théorie des Ensembles, Ch. I, II, Hermann, 1954

4 N. Bourbaki, *Eléments de Mathématique,* Première partie, Livre II, Algèbre, Ch. I, Hermann, 1942

5 N. Bourbaki, *Eléments de Mathématique,* Première partie, Livre II, Algèbre, Ch. VI, Hermann, 1952

6 N. Bourbaki, *Eléments de Mathématique,* Première partie, Livre III, Topologie générale, Ch. IV, Nombres réels, 2nd ed., Hermann, 1959

7 N. Bourbaki, *Eléments de Mathématique,* Deuxième partie, Livre II, Algèbre commutative, Ch. II, Hermann, 1952

8 L. W. Cohen and C. Goffman, A theory of transfinite convergence, *Transactions of the Amer. Math. Soc.,* Vol. 66, 1949

9 R. Dedekind, *Was sind und was sollen die Zahlen?* Vierweg, 1898

10 A. Fraenkel, *Abstract Set Theory,* Amsterdam, 1952

11 K. Gödel, Uber die formal unentscheidbaren Sätze der Principia Math. und verwandter Systeme I, *Monatsh. für Math. und Phys.*, Vol. 38, 1931

12 P. Halmos, *Naive Set Theory,* Van Nostrand, 1960

13 N. Jacobson, *Lectures in Abstract Algebra,* Vol. I, Van Nostrand, 1958

14 E. Landau, *Foundations of Analysis,* Chelsea, 1951

15 E. Nagel and J. R. Newman, *Gödel's Proof,* New York Univ. Press, 1958

16 G. Peano, *Notations de logique math.; intr. au Formulaire de Math.*, Turin, 1894

17 G. Peano, *Formulaire de Math.*, 1895–1905

18 L. Pontrjagin, *Topological Groups,* Ch. V, Princeton Univ. Press, 1939

19 B. Russell, *Introduction to Mathematical Philosophy,* Allen & Unwin, 1924

20 B. Russell and A. N. Whitehead, *The Principles of Mathematics,* 2nd ed., Norton, 1938

21 W. Sierpínski, *Leçons sur les nombres transfinis,* Gauthier-Villars, 1950

22 P. Suppes, *Axiomatic Set Theory,* Van Nostrand, 1960

23 B. L. Van der Waerden, *Modern Algebra,* Frederick Ungar Pub. Co., 1953

24 O. Zariski and P. Samuel, *Commutative Algebra,* Vol. I, Van Nostrand, 1958

# Index